NATURE NEAR LONDON

PREFACE

It is usually supposed to be necessary to go far into the country to find wild birds and animals in sufficient numbers to be pleasantly studied. Such was certainly my own impression till circumstances led me, for the convenience of access to London, to reside for awhile about twelve miles from town. There my preconceived views on the subject were quite overthrown by the presence of as much bird-life as I had been accustomed to in distant fields and woods.

First, as the spring began, came crowds of chiffchaffs and willow-wrens, filling the furze with ceaseless flutterings. Presently a nightingale sang in a hawthorn bush only just on the other side of the road. One morning, on looking out of window, there was a hen pheasant in the furze almost underneath. Rabbits often came out into the spaces of sward between the bushes.

The furze itself became a broad surface of gold, beautiful to look down upon, with islands of tenderest birch green interspersed, and willows in which the sedge-reedling chattered. They used to say in the country that cuckoos were getting scarce, but here the notes of the cuckoo echoed all day long, and the birds often flew over the house. Doves cooed, blackbirds whistled, thrushes sang, jays called, wood-pigeons uttered the old familiar notes in the little copse hard by. Even a heron went over now and then, and in the evening from the window I could hear partridges calling each other to roost.

Along the roads and lanes the quantity and variety of life in the hedges was really astonishing. Magpies, jays, woodpeckers—both green and pied—kestrels hovering overhead, sparrow-hawks darting over gateways, hares by the clover, weasels on the mounds, stoats at the edge of the corn. I missed but two birds, the corncrake and the grasshopper lark, and found these another season. Two squirrels one day ran along the palings and up into a guelder-rose tree in the garden. As for the finches and sparrows their number was past calculation. There was material for many years' observation, and finding myself so unexpectedly in the midst of these things, I was led to make the following sketches, which were published in *The Standard*, and are now reprinted by permission.

The question may be asked: Why have you not indicated in every case the precise locality where you were so pleased? Why not mention the exact hedge, the particular meadow? Because no two persons look at the same thing with the same eyes. To me this spot may be attractive, to you another; a third thinks yonder gnarled oak the most artistic. Nor could I guarantee that every one should see the same things under the same conditions of season, time, or weather. How could I arrange for you next autumn to see the sprays of the horse-chestnut, scarlet from frost, reflected in the dark water of the brook? There might not be any frost till all the leaves had dropped. How could I contrive that the cuckoos should circle round the copse, the sunlight glint upon the stream, the warm sweet wind come breathing over the young corn just when I should wish you to feel it? Every one must find their own locality. I find a favourite wild-flower here, and the spot is dear to me; you find yours yonder. Neither painter nor writer can show the spectator their originals. It would be very easy, too, to pass any of these places and see nothing, or but little. Birds are wayward, wild creatures uncertain. The tree crowded with wood-pigeons one minute is empty the next. To traverse the paths day by day, and week by week; to keep an eye ever on the fields from year's end to year's end, is the one only method of knowing what really is in or comes to them. That the sitting gambler sweeps the board is true of these matters. The richest locality may be apparently devoid of interest just at the juncture of a chance visit.

Though my preconceived ideas were overthrown by the presence of so much that was beautiful and interesting close to London, yet in course of time I came to understand what was at first a dim sense of something wanting. In the shadiest lane, in the still pinewoods, on the hills of purple heath, after brief contemplation there arose a restlessness, a feeling that it was essential to be moving. In no grassy mead was there a nook where I could stretch myself in slumberous ease and watch the swallows ever wheeling, wheeling in the sky. This was the unseen influence of mighty London. The strong life of the vast city magnetised me, and I felt it under the calm oaks. The something wanting in the fields was the absolute quiet, peace, and rest which dwells in the meadows and under the trees and on the hilltops in the country. Under its power the mind gradually yields itself to the green earth, the wind among the trees, the song of birds, and comes to have an understanding with them all. For this it is still necessary to seek the far-away glades and hollow coombes, or to sit alone beside the sea. That such a sense of quiet might not be lacking, I have added a chapter or so on those lovely downs that overlook the south coast.

R. J.

CONTENTS

WOODLANDS

The tiny white petals of the barren strawberry open under the April sunshine which, as yet unchecked by crowded foliage above, can reach the moist banks under the trees. It is then that the first stroll of the year should be taken in Claygate Lane. The slender runners of the strawberries trail over the mounds among the moss, some of the flowers but just above the black and brown leaves of last year which fill the shallow ditch. These will presently be hidden under the grass which is pushing up long blades, and bending over like a plume.

Crimson stalks and leaves of herb Robert stretch across the little cavities of the mound; lower, and rising almost from the water of the ditch, the wild parsnip spreads its broad fan. Slanting among the underwood, against which it leans, the dry white "gix" (cow-parsnip) of last year has rotted from its root, and is only upheld by branches.

Yellowish green cup-like leaves are forming upon the brown and drooping heads of the spurge, which, sheltered by the bushes, has endured the winter's frosts. The lads pull them off, and break the stems, to watch the white "milk" well up, the whole plant being full of acrid juice. Whorls of woodruff and grass-like leaves of stitchwort are rising; the latter holds but feebly to the earth, and even in snatching the flower the roots sometimes give way and the plant is lifted with it.

Upon either hand the mounds are so broad that they in places resemble covers rather than hedges, thickly grown with bramble and briar, hazel and hawthorn, above which the straight trunks of young oaks and Spanish chestnuts stand in crowded but careless ranks. The leaves which dropped in the preceding autumn from these trees still lie on the ground under the bushes, dry and brittle, and the blackbirds searching about among them cause as much rustling as if some animal were routing about.

As the month progresses these wide mounds become completely green, hawthorn and bramble, briar and hazel put forth their leaves, and the eye can no longer see into the recesses. But above, the oaks and edible chestnuts are still dark and leafless, almost black by contrast with the vivid green beneath them. Upon their bare boughs the birds are easily seen, but the moment they descend among the bushes are difficult to find. Chaffinches call and challenge continually—these trees are their favourite resort—and yellowhammers flit along the underwood.

Behind the broad hedge are the ploughed fields they love, alternating with meadows down whose hedges again a stream of birds is always flowing to the lane. Bright as are the colours of the yellowhammer, when he alights among the brown clods of the ploughed field he is barely visible, for brown conceals like vapour. A white butterfly comes fluttering along the lane, and as it passes under a tree a chaffinch swoops down and snaps at it, but rises again without doing apparent injury, for the butterfly continues its flight.

From an oak overhead comes the sweet slender voice of a linnet, the sunshine falling on his rosy breast. The gateways show the thickness of the hedge, as an embrasure shows the thickness of a wall. One gives entrance to an arable field which has been recently rolled, and along the gentle rise of a "land" a cock-pheasant walks, so near that the ring about his neck is visible. Presently, becoming conscious that he is observed, he goes down into a furrow, and is then hidden.

The next gateway, equally deep-set between the bushes, opens on a pasture, where the docks of last year still cumber the ground, and bunches of rough grass and rushes are scattered here and there. A partridge separated from his mate is calling across the field, and comes running over the short sward as his companion answers. With his neck held high and upright, stretched to see around, he looks larger than would be supposed, as he runs swiftly, threading his way through the tufts, the docks, and the rushes. But suddenly noticing that the gateway is not clear, he crouches, and is concealed by the grass.

Some distance farther there is a stile, sitting upon which the view ranges over two adjacent meadows. They are bounded by a copse of ash stoles and young oak trees, and the lesser of the meads is full of rush bunches and dotted with green ant-hills. Among these, just beyond gunshot, two rabbits are feeding; pausing and nibbling till they have eaten the tenderest blades, and then leisurely hopping a yard or so to another spot. Later on in the summer this little meadow which divides the lane from the copse is alive with rabbits.

Along the hedge the brake fern has then grown, in the corner by the copse there is a beautiful mass of it, and several detached bunches away from the hedge among the ant-hills. From out of the fern, which is a favourite retreat with them, rabbits are continually coming, feeding awhile, darting after each other, and back again to cover. To-day there are but three, and they do not venture far from their buries.

Watching these, a green woodpecker cries in the copse, and immediately afterwards flies across the mead, and away to another plantation. Occasionally the spotted woodpecker may be seen here, a little bird which, in the height of summer, is lost among the foliage, but in spring and winter can be observed tapping at the branches of the trees.

I think I have seen more spotted woodpeckers near London than in far distant and nominally wilder districts. This lane, for some two miles, is lined on each side with trees, and, besides this particular copse, there are several others close by; indeed, stretching across the country to another road, there is a succession of copses, with meadows between. Birds which love trees are naturally seen flitting to and fro in the lane; the trees are at present young, but as they grow older and decay they will be still more resorted to.

Jays screech in the trees of the lane almost all the year round, though more frequently in spring and autumn, but I rarely walked here without seeing or hearing one. Beyond the stile, the lane descends into a hollow, and is bordered by a small furze common, where, under shelter of the hollow brambles and beneath the golden bloom of the furze, the pale anemones flower.

When the June roses open their petals on the briars, and the scent of new-mown hay is wafted over the hedge from the meadows, the lane seems to wind through a continuous wood. The oaks and chestnuts, though too young to form a complete arch, cross their green branches, and cast a delicious shadow. For it is in the shadow that we enjoy the summer, looking forth from the gateway upon the mowing grass where the glowing sun pours down his fiercest beams.

Tall bennets and red sorrel rise above the grass, white ox-eye daisies chequer it below; the distant hedge quivers as the air, set in motion by the intense heat, runs along. The sweet murmuring coo of the turtle dove comes from the copse, and the rich notes of the blackbird from the oak into which he has mounted to deliver them.

Slight movements in the hawthorn, or in the depths of the tall hedge grasses, movements too quick for the glance to catch their cause, are where some tiny bird is passing from spray to spray. It may be a white-throat creeping among the nettles after his wont, or a wren. The spot where he was but a second since may be traced by the trembling of the leaves, but the keenest attention may fail to detect where he is now. That slight motion in the hedge, however, conveys an impression of something living everywhere within.

There are birds in the oaks overhead whose voice is audible though they are themselves unseen. From out of the mowing grass, finches rise and fly to the hedge; from the hedge again others fly out, and, descending into the grass, are concealed as in a forest. A thrush travelling along the hedgerow just outside goes by the gateway within a yard. Bees come upon the light wind, gliding with it, but with their bodies aslant across the line of current. Butterflies flutter over the mowing grass, hardly clearing the bennets. Many-coloured insects creep up the sorrel stems and take wing from the summit.

Everything gives forth a sound of life. The twittering of swallows from above, the song of greenfinches in the trees, the rustle of hawthorn sprays moving under the weight of tiny creatures, the buzz upon the breeze; the very flutter of the butterflies' wings, noiseless as it is, and the wavy movement of the heated air across the field cause a sense of motion and of music.

The leaves are enlarging, and the sap rising, and the hard trunks of the trees swelling with its flow; the grass blades pushing upwards; the seeds completing their shape; the tinted petals uncurling. Dreamily listening, leaning on the gate, all these are audible to the inner senses, while the ear follows the midsummer hum, now sinking, now sonorously increasing over the oaks. An effulgence fills the southern boughs, which the eye cannot sustain, but which it knows is there.

The sun at its meridian pours forth his light, forgetting, in all the inspiration of his strength and glory, that without an altar-screen of green his love must scorch. Joy in life; joy in life. The ears listen, and want more: the eyes are gratified with gazing, and desire yet further; the nostrils are filled with the sweet odours of flower and sap. The touch, too, has its pleasures, dallying with leaf and flower. Can you not almost grasp the odour-laden air and hold it in the hollow of the hand?

Leaving the spot at last, and turning again into the lane, the shadows dance upon the white dust under the feet, irregularly circular spots of light surrounded with umbra shift with the shifting branches. By the wayside lie rings of dandelion stalks carelessly cast down by the child who made them, and tufts of delicate grasses gathered for their beauty but now sprinkled with dust. Wisps of hay hang from the lower boughs of the oaks where they brushed against the passing load.

After a time, when the corn is ripening, the herb betony flowers on the mounds under the oaks. Following the lane down the hill and across the small furze common at the bottom, the marks of traffic fade away, the dust ceases, and is succeeded by sward. The hedgerows on either side are here higher than ever, and are thickly fringed with bramble bushes, which sometimes encroach on the waggon ruts in the middle, and are covered with flowers, and red, and green, and ripe blackberries together.

Green rushes line the way, and green dragon flies dart above them. Thistledown is pouting forth from the swollen tops of thistles crowded with seed. In a gateway the turf has been worn away by waggon wheels and the hoofs of cart horses, and the dry heat has pulverised the crumbling ruts. Three hen pheasants and a covey of partridges that have been dusting themselves here move away without much haste at the approach of footsteps—the pheasants into the thickets, and the partridges through the gateway. The shallow holes in which they were sitting can be traced on the dust, and there are a few small feathers lying about.

A barley field is within the gate; the mowers have just begun to cut it on the opposite side. Next to it is a wheat field; the wheat has been cut and stands in shocks. From the stubble by the nearest shock two turtle doves rise, alarmed, and swiftly fly towards a wood which bounds the field. This wood, indeed, upon looking again, clearly bounds not this field only, but the second and the third, and so far as the eye can see over the low hedges of the corn, the trees continue. The green lane as it enters the wood, becomes wilder and rougher at every step, widening, too, considerably.

In the centre the wheels of timber carriages, heavily laden with trunks of trees which were dragged through by straining teams in the rainy days of spring, have left vast ruts, showing that they must have sunk to the axle in the soft clay. These then filled with water, and on the water duck-weed grew, and aquatic grasses at the sides. Summer heats have evaporated the water, leaving the weeds and grasses prone upon the still moist earth.

Rushes have sprung up and mark the line of the ruts, and willow stoles, bramble bushes, and thorns growing at the side, make, as it were, a third hedge in the middle of the lane. The best path is by the wood itself, but even there occasional leaps are necessary over pools of dark water full of vegetation. These alternate with places where the ground, being higher, yawns with wide cracks crumbling at the edge, the heat causing the clay to split and open. In winter it must be an impassable quagmire; now it is dry and arid.

Rising out of this low-lying spot the lane again becomes green and pleasant, and is crossed by another. At the meeting of these four ways some boughs hang over a green bank where I have often rested. In front the lane is barred by a gate, but beyond the gate it still continues its straight course into the wood. To the left the track, crossing at right angles, also proceeds into the wood, but it is so overhung with trees and blocked by bushes that its course after the first hundred yards or so cannot be traced.

To the right the track—a little wider and clearer of bushes—extends through wood, and as it is straight and rises up a gentle slope, the eye can travel along it half a mile. There is nothing but wood around. This track to the right appears the most used, and has some ruts in the centre. The sward each side is concealed by endless thistles, on the point of sending forth clouds of thistledown, and to which presently the goldfinches will be attracted.

Occasionally a movement among the thistles betrays the presence of a rabbit; only occasionally, for though the banks are drilled with buries, the lane is too hot for them at midday. Particles of rabbits' fur lie on the ground, and their runs are visible in every direction. But there are no birds. A solitary robin, indeed, perches on an ash branch opposite, and regards me thoughtfully. It is impossible to go anywhere in the open air without a robin; they are the very spies of the wood. But there are no thrushes, no blackbirds, finches, nor even sparrows.

In August it is true most birds cease to sing, but sitting thus partially hidden and quiet, if there were any about something would be heard of them. There would be a rustling, a thrush would fly across the lane, a blackbird would appear by the gateway yonder in the shadow which he loves, a finch would settle in the oaks. None of these incidents occur; none of the lesser signs of life in the foliage, the tremulous spray, the tap of a bill cleaned by striking first one side and then the other against a bough, the rustle of a wing—nothing.

There are woods, woods, woods; but no birds. Yonder a drive goes straight into the ashpoles, it is green above and green below, but a long watch will reveal nothing living. The dry mounds must be full of rabbits, there must be pheasants somewhere; but nothing visible. Once only a whistling sound in the air directs the glance upwards, it is a wood-pigeon flying at full speed. There are no bees, for there are no flowers. There are no butterflies. The black flies are not numerous, and rarely require a fanning from the ash spray carried to drive them off.

Two large dragon-flies rush up and down, and cross the lane, and rising suddenly almost to the tops of the oaks swoop down again in bold sweeping curves. The broad, deep ditch between the lane and the mound of the wood is dry, but there are no short rustling sounds of mice.

The only sound is the continuous singing of the grasshoppers, and the peculiar snapping noise they make as they spring, leaping along the sward. The fierce sun of the ripe wheat pours down a fiery glow scarcely to be borne except under the boughs; the hazel leaves already have lost their green, the tips of the rushes are shrivelling, the grass becoming brown; it is a scorched and parched desert of wood.

The finches have gone forth in troops to the stubble where the wheat has been cut, and where they can revel on the seeds of the weeds now ripe. Thrushes and blackbirds have gone to the streams, to splash and bathe, and to the mown meadows, where in the short aftermath they can find their food. There they will look out on the shady side of the hedge as the sun declines, six or eight perhaps of them along the same hedge, but all in the shadow, where the dew forms first as the evening falls, where the grass feels cool and moist, while still on the sunny side it is warm and dry.

The bees are busy on the heaths and along the hilltops, where there are still flowers and honey, and the butterflies are with them. So the woods are silent, still, and deserted, save by a stray rabbit among the thistles, and the grasshoppers ceaselessly leaping in the grass.

Returning presently to the gateway just outside the wood, where upon first coming the pheasants and partridges were dusting themselves, a waggon is now passing among the corn and is being laden with the sheaves. But afar off, across the broad field and under the wood, it seems somehow only a part of the silence and the solitude. The men with it move about the stubble, calmly toiling; the horses, having drawn it a little way, become motionless, reposing as they stand, every line of their large limbs expressing delight in physical ease and idleness.

Perhaps the heat has made the men silent, for scarcely a word is spoken; if it were, in the stillness it must be heard, though they are at some distance. The wheels, well greased for the heavy harvest work, do not creak. Save an occasional monosyllable, as the horses are ordered on, or to stop, and a faint rustling of straw, there is no sound. It may be the flood of brilliant light, or the mirage of the heat, but in some way the waggon and its rising load, the men and the horses, have an unreality of appearance.

The yellow wheat and stubble, the dull yellow of the waggon, toned down by years of weather, the green woods near at hand, darkening in the distance and slowly changing to blue, the cloudless sky, the heat-suffused atmosphere, in which things seem to float rather than to grow or stand, the shadowless field, all are there, and yet are not there, but far away and vision-like. The waggon, at last laden, travels away, and seems rather to disappear of itself than to be hidden by the trees. It is an effort to awake and move from the spot.

FOOTPATHS

"Always get over a stile," is the one rule that should ever be borne in mind by those who wish to see the land as it really is—that is to say, never omit to explore a footpath, for never was there a footpath yet which did not pass something of interest.

In the meadows, everything comes pressing lovingly up to the path. The small-leaved clover can scarce be driven back by frequent footsteps from endeavouring to cover the bare earth of the centre. Tall buttercups, round whose stalks the cattle have carefully grazed, stand in ranks; strong ox-eye daisies, with broad white disks and torn leaves, form with the grass the tricolour of the pasture—white, green, and gold.

When the path enters the mowing grass, ripe for the scythe, the simplicity of these cardinal hues is lost in the multitude of shades and the addition of other colours. The surface of mowing grass is indeed made up of so many tints that at the first glance it is confusing; and hence, perhaps, it is that hardly ever has an artist succeeded in getting the effect upon canvas. Of the million blades of grass no two are of the same shade.

Pluck a handful and spread them out side by side and this is at once evident. Nor is any single blade the same shade all the way up. There may be a faint yellow towards the root, a full green about the middle, at the tip perhaps the hot sun has scorched it, and there is a trace of brown. The older grass, which comes up earliest, is distinctly different in tint from that which has but just reached its greatest height, and in which the sap has not yet stood still.

Under all there is the new grass, short, sweet, and verdant, springing up fresh between the old, and giving a tone to the rest as you look down into the bunches. Some blades are nearly grey, some the palest green, and among them others, torn from the roots perhaps by rooks searching for grubs, are quite white. The very track of a rook through the grass leaves a different shade each side, as the blades are bent or trampled down.

The stalks of the bennets vary, some green, some yellowish, some brown, some approaching whiteness, according to age and the condition of the sap. Their tops, too, are never the same, whether the pollen clings to the surface or whether it has gone. Here the green is almost lost in red, or quite; here the grass has a soft, velvety look; yonder it is hard and wiry, and again graceful and drooping. Here there are bunches so rankly verdant that no flower is visible and no other tint but dark green; here it is thin and short, and the flowers, and almost the turf itself, can be seen; then there is an array of bennets (stalks which bear the grass-seed) with scarcely any grass proper.

Every variety of grass—and they are many—has its own colour, and every blade of every variety has its individual variations of that colour. The rain falls, and there is a darker tint at large upon the field, fresh but darker; the sun shines and at first the hue is lighter, but presently if the heat last a brown comes. The wind blows, and immediately as the waves of grass roll across the meadow a paler tint follows it.

A clouded sky dulls the herbage, a cloudless heaven brightens it, so that the grass almost reflects the firmament like water. At sunset the rosy rays bring out every tint of red or purple. At noonday, watch as alternate shadow and sunshine come one after the other as the clouds are wafted over. By moonlight perhaps the white ox-eyed daisies show the most. But never will you find the mowing grass in the same field looking twice alike.

Come again the day after to-morrow only, and there is a change; some of the grass is riper, some is thicker, with further blades which have pushed up, some browner. Cold northern winds cause it to wear a dry, withered aspect; under warm showers it visibly opens itself; in a hurricane it tosses itself wildly to and fro; it laughs under the sunshine.

There are thick bunches by the footpath, which hang over and brush the feet. While approaching there seems nothing there except grass, but in the act of passing, and thus looking straight down into them, there are blue eyes at the bottom gazing up. These specks of blue sky hidden in the grass tempt the hand to gather them, but then you cannot gather the whole field.

Behind the bunches where the grass is thinner are the heads of purple clover; pluck one of these, and while meditating draw forth petal after petal and imbibe the honey with the lips till nothing remains but the green framework, like stolen jewellery from which the gems have been taken. Torn pink ragged robins through whose petals a comb seems to have been remorselessly dragged, blue scabious, red knapweeds, yellow rattles, yellow vetchings by the hedge, white flowering parsley, white campions, yellow tormentil, golden buttercups, white cuckoo-flowers, dandelions, yarrow, and so on, all carelessly sown broadcast without order or method, just as negligently as they are named here, first remembered, first mentioned, and many forgotten.

Highest and coarsest of texture, the red-tipped sorrel—a crumbling red—so thick and plentiful that at sunset the whole mead becomes reddened. If these were in any way set in order or design, howsoever entangled, the eye might, as it were, get at them for reproduction. But just where there should be flowers there are none, whilst in odd places where there are none required there are plenty.

In hollows, out of sight till stumbled on, is a mass of colour; on the higher foreground only a dull brownish green. Walk all round the meadow, and still no vantage point can be found where the herbage groups itself, whence a scheme of colour is perceivable. There is no "artistic" arrangement anywhere.

So, too, with the colours—of the shades of green something has already been said—and here are bright blues and bright greens, yellows

24

and pinks, positive discords and absolute antagonisms of tint side by side, yet without jarring the eye. Green all round, the trees and hedges; blue overhead, the sky; purple and gold westward, where the sun sinks. No part of this grass can be represented by a blur or broad streak of colour, for it is not made up of broad streaks. It is composed of innumerable items of grass blade and flower, each in itself coloured and different from its neighbour. Not one of these must be slurred over if you wish to get the same effect.

Then there are drifting specks of colour which cannot be fixed. Butterflies, white, parti-coloured, brown, and spotted, and light blue flutter along beside the footpath; two white ones wheel about each other, rising higher at every turn till they are lost and no more to be distinguished against a shining white cloud. Large dark humble bees roam slowly, and honey bees with more decided flight. Glistening beetles, green and gold, run across the bare earth of the path, coming from one crack in the dry ground and disappearing in the (to them) mighty chasm of another.

Tiny green "hoppers"—odd creatures shaped something like the fancy frogs of children's story-books—alight upon it after a spring, and pausing a second, with another toss themselves as high as the highest bennet (veritable elm-trees by comparison), to fall anywhere out of sight in the grass. Reddish ants hurry over. Time is money; and their business brooks no delay.

Bee-like flies of many stripes and parti-coloured robes face you, suspended in the air with wings vibrating so swiftly as to be unseen; then suddenly jerk themselves a few yards to recommence hovering. A greenfinch rises with a yellow gleam and a sweet note from the grass, and is off with something for his brood, or a starling, solitary now, for his mate is in the nest, startled from his questing, goes straight away.

Dark starlings, greenfinch, gilded fly, glistening beetle, blue butterfly, humble bee with scarf about his thick waist, add their moving dots of colour to the surface. There is no design, no balance, nothing like a pattern perfect on the right-hand side, and exactly equal on the left-hand. Even

25

trees which have some semblance of balance in form are not really so, and as you walk round them so their outline changes.

Now the path approaches a stile set deep in thorns and brambles, and hardly to be gained for curved hooks and prickles. But on the briars June roses bloom, arches of flowers over nettles, burdock, and rushes in the ditch beneath. Sweet roses—buds yet unrolled, white and conical; roses half open and pink tinted; roses widespread, the petals curling backwards on the hedge, abandoning their beauty to the sun. In the pasture over the stile a roan cow feeds unmoved, calmly content, gathering the grass with rough tongue. It is not only what you actually see along the path, but what you remember to have seen, that gives it its beauty.

From hence the path skirts the hedge enclosing a copse, part of which had been cut in the winter, so that a few weeks since in spring the bluebells could be seen, instead of being concealed by the ash branches and the woodbine. Among them grew one with white bells, like a lily, solitary in the midst of the azure throng. A "drive," or green lane passing between the ash-stoles, went into the copse, with tufts of tussocky grass on either side and rush bunches, till farther away the overhanging branches, where the poles were uncut, hid its course.

Already the grass has hidden the ruts left by the timber carriages— the last came by on May-day with ribbons of orange, red, and blue on the horses' heads for honour of the day. Another, which went past in the wintry weeks of the early year, was drawn by a team wearing the ancient harness with bells under high hoods, or belfries, bells well attuned, too, and not far inferior to those rung by handbell men. The beat of the three horses' hoofs sounds like the drum that marks time to the chime upon their backs. Seldom, even in the far away country, can that pleasant chime be heard.

But now the timber is all gone, the ruts are hidden, and the tall spruce firs, whose graceful branches were then almost yellow with young needles on the tip, are now clothed in fresh green. On the bank there is a flower which is often gathered for the forget-me-not, and is not unlike it at the first glance; but if the two be placed side by side, this, the scorpion grass, is but a pale imitation of the true plant; its petals vary in colour and are often dull, and it has not the yellow central spot. Yet it is not unfrequently sold in pots in the shops as forget-me-not. It flowers on the bank, high above the water of the ditch.

The true forget-me-not can hardly be seen in passing, so much does it nestle under flags and behind sedges, and it is not easy to gather because it flowers on the very verge of the running stream. The shore is bordered with matted vegetation, aquatic grass, and flags and weeds, and outside these, where its leaves are washed and purified by the clear stream, its blue petals open. Be cautious, therefore, in reaching for the forget-me-not, lest the bank be treacherous.

It was near this copse that in early spring I stayed to gather some white sweet violets, for the true wild violet is very nearly white. I stood close to a hedger and ditcher, who, standing on a board, was cleaning out the mud that the water might run freely. He went on with his work, taking not the least notice of an idler, but intent upon his labour, as a good and true man should be. But when I spoke to him he answered me in clear, well-chosen language, well pronounced, "in good set terms."

No slurring of consonants and broadening of vowels, no involved and backward construction depending on the listener's previous knowledge for comprehension, no half sentences indicating rather than explaining, but correct sentences. With his shoes almost covered by the muddy water, his hands black and grimy, his brown face splashed with mud, leaning on his shovel he stood and talked from the deep ditch, not much more than head and shoulders visible above it. It seemed a voice from the very earth, speaking of education, change, and possibilities.

The copse is now filling up with undergrowth; the brambles are spreading, the briars extending, masses of nettles, and thistles like saplings in size and height, crowding the spaces between the ash-stoles. By the banks great cow-parsnips or "gix" have opened their broad heads of white flowers; teazles have lifted themselves into view, every opening is occupied. There is a scent of elder flowers, the meadow-sweet is pushing up, and will soon be out, and an odour of new-mown hay floats on the breeze.

From the oak green caterpillars slide down threads of their own making to the bushes below, but they are running terrible risk. For a pair of white-throats or "nettle-creepers" are on the watch, and seize the green creeping things crossways in their beaks. Then they perch on a branch three or four yards only from where I stand, silent and motionless, and glance first at me and next at a bush of bramble which projects out to the edge of the footpath. So long as my eyes are turned aside, or half closed, the bird perches on the branch, gaining confidence every moment. The instant I open my eyes, or move them, or glance towards him, without either movement of head, hand, or foot, he is off to the oak.

His tiny eyes are intent on mine; the moment he catches my glance he retires. But in half a minute affection brings him back, still with the caterpillar in his beak, to the same branch. Whilst I have patience to look the other way there he stays, but again a glance sends him away. This is repeated four or five times, till, finally, convinced that I mean no harm and yet timorous and fearful of betrayal even in the act, he dives down into the bramble bush.

After a brief interval he reappears on the other side of it, having travelled through and left his prey with his brood in the nest there. Assured by his success his mate follows now, and once having done it, they continue to bring caterpillars, apparently as fast as they can pass between the trees and the bush. They always enter the bush, which is scarcely two yards from me, on one side, pass through in the same direction, and emerge on the other side, having thus regular places of entrance and exit.

As I stand watching these birds a flock of rooks goes over, they have left the nesting trees, and fly together again. Perhaps this custom of nesting together in adjacent trees and using the same one year after year is not so free from cares and jealousies as the solitary plan of the little white-throats here. Last March I was standing near a rookery, noting the contention and quarrelling, the downright tyranny, and brigandage which is carried on there. The very sound of the cawing, sharp and angry, conveys the impression of hate and envy.

Two rooks in succession flew to a nest the owners of which were absent, and deliberately picked a great part of it to pieces, taking the twigs for their own use. Unless the rook, therefore, be ever in his castle his labour is torn down, and, as with men in the fierce struggle for wealth, the meanest advantages are seized on. So strong is the rook's bill that he tears living twigs of some size with it from the bough. The white-throats were without such envy and contention.

From hence the footpath, leaving the copse, descends into a hollow, with a streamlet flowing through a little meadow, barely an acre, with a pollard oak in the centre, the rising ground on two sides shutting out all but the sky, and on the third another wood. Such a dreamy hollow might be painted for a glade in the Forest of Arden, and there on the sward and leaning against the ancient oak one might read the play through without being disturbed by a single passer-by. A few steps farther and the stile opens on a road.

There the teams travel with rows of brazen spangles down their necks, some with a wheatsheaf for design, some with a swan. The road itself, if you follow it, dips into a valley where the horses must splash through the water of a brook spread out some fifteen or twenty yards wide; for, after the primitive Surrey fashion, there is no bridge for waggons. A narrow wooden structure bears foot-passengers; you cannot but linger half across and look down into its clear stream. Up the current where it issues from the fields and falls over a slight obstacle the sunlight plays and glances.

A great hawthorn bush grows on the bank; in spring, white with May; in autumn, red with haws or peggles. To the shallow shore of the brook, where it washes the flints and moistens the dust, the house-martins come for mortar. A constant succession of birds arrive all day long to drink at the clear stream, often alighting on the fragments of chalk and flint which stand in the water, and are to them as rocks.

Another footpath leads from the road across the meadows to where the brook is spanned by the strangest bridge, built of brick, with one arch, but only just wide enough for a single person to walk, and with parapets only four or five inches high. It is thrown aslant the stream, and not straight across it, and has a long brick approach. It is not unlike—on a small scale—the bridges seen in views of Eastern travel. Another path leads to a hamlet, consisting of a church, a farmhouse, and three or four cottages—a veritable hamlet in every sense of the word.

In a village a few miles distant, as you walk between cherry and pear orchards, you pass a little shop—the sweets, and twine, and trifles are such as may be seen in similar windows a hundred miles distant. There is the very wooden measure for nuts, which has been used time out of mind, in the distant country. Out again into the road as the sun sinks, and westwards the wind lifts a cloud of dust, which is lit up and made rosy by the rays passing through it. For such is the beauty of the sunlight that it can impart a glory even to dust.

Once more, never go by a stile (that does not look private) without getting over it and following the path. But they all end in one place. After rambling across furze and heath, or through dark fir woods; after lingering in the meadows among the buttercups, or by the copses where the pheasants crow; after gathering June roses, or, in later days, staining the lips with blackberries or cracking nuts, by-and-by the path brings you in sight of a railway station. And the railway station, through some process of mind, presently compels you to go up on the platform, and after a little puffing and revolution of wheels you emerge at Charing Cross, or London Bridge, or Waterloo, or Ludgate Hill, and, with the freshness of the meadows still clinging to your coat, mingle with the crowd.

The inevitable end of every footpath round about London is London. All paths go thither.

If it were far away in the distant country you might sit down in the shadow upon the hay and fall asleep, or dream awake hour after hour. There would be no inclination to move. But if you sat down on the sward under the ancient pollard oak in the little mead with the brook, and the wood of which I spoke just now as like a glade in the enchanted Forest of Arden, this would not be possible. It is the proximity of the immense City which induces a mental, a nerve-restlessness. As you sit and would dream a something plucks at the mind with constant reminder; you cannot dream for long, you must up and away, and, turn in which direction you please, ultimately it will lead you to London.

There is a fascination in it; there is a magnetism stronger than that of the rock which drew the nails from Sindbad's ship. You are like a bird let out with a string tied to the foot to flutter a little way and return again. It is not business, for you may have none, in the ordinary sense; it is not "society," it is not pleasure. It is the presence of man in his myriads. There is something in the heart which cannot be satisfied away from it.

It is a curious thing that your next-door neighbour may be a stranger, but there are no strangers in a vast crowd. They all seem to have some relationship, or rather, perhaps, they do not rouse the sense of reserve which a single unknown person might. Still, the impulse is not to be analysed; these are mere notes acknowledging its power. The hills and vales, and meads and woods are like the ocean upon which Sindbad sailed; but coming too near the loadstone of London, the ship wends thither, whether or no.

At least it is so with me, and I often go to London without any object whatever, but just because I must, and, arriving there, wander whithersoever the hurrying throng carries me.

FLOCKS OF BIRDS

A certain road leading outwards from a suburb, enters at once among fields. It soon passes a thick hedge dividing a meadow from a cornfield, in which hedge is a spot where some bluebells may be found in spring. Wild flowers are best seen when in masses, a few scattered along a bank much concealed by grass and foliage are lost, except indeed, upon those who love them for their own sake.

This meadow in June, for instance, when the buttercups are high, is one broad expanse of burnished gold. The most careless passer-by can hardly fail to cast a glance over acres of rich yellow. The furze, again, especially after a shower has refreshed its tint, must be seen by all. Where broom grows thickly, lifting its colour well into view, or where the bird's-foot lotus in full summer overruns the thin grass of some upland pasture, the eye cannot choose but acknowledge it. So, too, with charlock, and with hill sides purple with heath, or where the woodlands are azure with bluebells for a hundred yards together. Learning from this, those who would transplant wild flowers to their garden should arrange to have as many as possible of the same species close together.

The bluebells in this hedge are unseen, except by the rabbits. The latter have a large burrow, and until the grass is too tall, or after it is cut or grazed, can be watched from the highway. In this hedge the first nightingale of the year sings, beginning some two or three days before the bird which comes to the bushes in the gorse, which will presently be mentioned.

It is, or rather was, a favourite meadow with the partridges; one summer there was, I think, a nest in or near it, for I saw the birds there daily. But the next year they were absent. One afternoon a brace of partridges came over the hedge within a few inches of my head; they had been flushed and frightened at some distance, and came with the wind at a tremendous pace. It is a habit with partridges to fly low, but just skimming the tops of the hedges, and certainly, had they been three inches lower, they must have taken my hat off. The knowledge that partridges were often about there, made me always glance into this field on passing it, long after the nesting season was over.

In October, as I looked as usual, a hawk flew between the elms, and out into the centre of the meadow, with a large object in his talons. He alighted in the middle, so as to be as far as possible from either hedge, and no doubt prepared to enjoy his quarry, when something startled him, and he rose again. Then, as I got a better view, I saw it was a rat he was carrying. The long body of the animal was distinctly visible, and the tail depending, the hawk had it by the shoulders or head. Flying without the least apparent effort, the bird cleared the elms, and I lost sight of him beyond them. Now, the kestrel is but a small bird, and taking into consideration the size of the bird, and the weight of a rat, it seems as great a feat in proportion as for an eagle to snatch up a lamb.

Some distance up the road, and in the corner of an arable field, there was a wheat rick which was threshed and most of the straw carted away. But there still remained the litter, and among it probably a quantity of stray corn. There was always a flock of sparrows on this litter—a flock that might often be counted by the hundred. As I came near the spot one day a sparrow-hawk, whose approach I had not observed, and which had therefore been flying low, suddenly came over the hedge just by the loose straw.

With shrill cries the sparrows instantly rushed for the hedge, not two yards distant; but the hawk, dashing through the crowd of them as they rose, carried away a victim. It was done in the tenth of a second. He came, singled his bird, and was gone like the wind, before the whirr of wings had ceased on the hawthorn where the flock cowered.

Another time, but in a different direction, I saw a hawk descend and either enter, or appear to enter, a short much-cropped hedge, but twenty yards distant. I ran to the spot; the hawk of course made off, but there was nothing in the bush save a hedge sparrow, which had probably attracted him, but which he had not succeeded in getting.

Kestrels are almost common; I have constantly seen them while strolling along the road, generally two together, and once three. In the latter part of the summer and autumn they seem to be most numerous, hovering over the recently reaped fields. Certainly there is no scarcity of hawks here. Upon one occasion, on Surbiton Hill, I saw a large bird of the same kind, but not sufficiently near to identify. From the gliding flight, the long forked tail, and large size I supposed it to be a kite. The same bird was going about next day, but still farther off. I cannot say that it was a kite, for unless it is a usual haunt, it is not in my opinion wise to positively identify a bird seen for so short a time.

The thick hedge mentioned is a favourite resort of blackbirds, and on a warm May morning, after a shower—they are extremely fond of a shower—half-a-dozen may be heard at once whistling in the elms. They use the elms here because there are not many oaks; the oak is the blackbird's favourite song-tree. There was one one day whistling with all his might on the lower branch of an elm, at the very roadside, and just above him a wood-pigeon was perched. A pair of turtle-doves built in the same hedge one spring, and while resting on the gate by the roadside their "coo-coo" mingled with the song of the nightingale and thrush, the blackbird's whistle, the chiff-chaff's "chip-chip," the willow-wren's pleading voice, and the rustle of green corn as the wind came rushing (as it always does to a gateway).

Goldfinches come by occasionally, not often, but still they do come. The rarest bird seems to be the bullfinch. I have only seen bullfinches three or four times in three seasons, and then only a pair. Now, this is worthy a note, as illustrating what I have often ventured to say about the habitat of birds being so often local, for if judged by observation here

34

the bullfinch would be said to be a scarce bird by London. But it has been stated upon the best authority that only a few miles distant, and still nearer town, they are common.

The road now becomes bordered by elms on either side, forming an irregular avenue. Almost every elm in spring has its chaffinch loudly challenging. The birdcatchers are aware that it is a frequented resort, and on Sunday mornings four or five of them used to be seen in the course of a mile, each with a call bird in a partly darkened cage, a stuffed dummy, and limed twigs. In the cornfields on either hand wood-pigeons are numerous in spring and autumn. Up to April they come in flocks, feeding on the newly sown grain when they can get at it, and varying it with ivy berries, from the ivy growing up the elms. By degrees the flocks break up as the nesting begins in earnest.

Some pair and build much earlier than others; in fact, the first egg recorded is very little to be depended on as an indication. Particular pairs (of many kinds of birds) may have nests, and yet the species as a species may be still flying in large packs. The flocks which settle in these fields number from one to two hundred. Rooks, wood-pigeons, and tame white pigeons often feed amicably mixed up together; the white tame birds are conspicuous at a long distance before the crops have risen, or after the stubble is ploughed.

I should think that the corn farmers of Surrey lose more grain from the birds than the agriculturists whose tenancies are a hundred miles from London. In the comparatively wild or open districts to which I had been accustomed before I made these observations I cannot recollect ever seeing such vast numbers of birds. There were places, of course, where they were numerous, and there were several kinds more represented than is the case here, and some that are scarcely represented at all. I have seen flocks of wood-pigeons immensely larger than any here; but then it was only occasionally. They came, passed over, and were gone. Here the flocks, though not very numerous, seem always to be about.

Sparrows crowd every hedge and field, their numbers are incredible; chaffinches are not to be counted; of greenfinches there must be thousands. From the railway even you can see them. I caught glimpses of a ploughed field recently sown one spring from the window of a railway carriage, every little clod of which seemed alive with small birds, principally sparrows, chaffinches, and greenfinches. There must have been thousands in that field alone. In autumn the numbers are even greater, or rather more apparent.

One autumn some correspondence appeared lamenting the scarcity of small birds (and again in the spring the same cry was raised); people said that they had walked along the roads or footpaths and there were none in the hedges. They were quite correct—the birds were not in the hedges, they were in the corn and stubble. After the nesting is well over and the wheat is ripe the birds leave the hedges and go out into the wheatfields; at the same time the sparrows quit the house-tops and gardens and do the same. At the very time this complaint was raised, the stubbles in Surrey, as I can vouch, were crowded with small birds.

If you walked across the stubble flocks of hundreds rose out of your way; if you leant on a gate and watched a few minutes you could see small flocks in every quarter of the field rising and settling again. These movements indicated a larger number in the stubble there, for where a great flock is feeding some few every now and then fly up restlessly. Earlier than that in the summer there was not a wheatfield where you could not find numerous wheatears picked as clean as if threshed where they stood. In some places, the wheat was quite thinned.

Later in the year there seems a movement of small birds from the lower to the higher lands. One December day I remember particularly visiting the neighbourhood of Ewell, where the lands begin to rise up towards the Downs. Certainly, I have seldom seen such vast numbers of small birds. Up from the stubble flew sparrows, chaffinches, greenfinches, yellow-hammers, in such flocks that the low-cropped hedge was covered with them. A second correspondence appeared in the spring upon the same subject, and again the scarcity of small birds was deplored.

So far as the neighbourhood of London was concerned, this was the exact reverse of the truth.

Small birds swarmed, as I have already stated, in every ploughed field. All the birdcatchers in London with traps and nets and limed twigs could never make the slightest appreciable difference to such flocks. I have always expressed my detestation of the birdcatcher; but it is founded on other grounds, and not from any fear of the diminution of numbers only. Where the birdcatcher does inflict irretrievable injury is in this way—a bird, say a nightingale, say a goldfinch, has had a nest for years in the corner of a garden, or an apple-tree in an orchard. The birdcatcher presently decoys one or other of these, and thenceforward the spot is deserted. The song is heard no more; the nest never again rebuilt.

The first spring I resided in Surrey I was fairly astonished and delighted at the bird life which proclaimed itself everywhere. The bevies of chiffchaffs and willow wrens which came to the thickets in the furze, the chorus of thrushes and blackbirds, the chaffinches in the elms, the greenfinches in the hedges, wood-pigeons and turtle-doves in the copses, tree-pipits about the oaks in the cornfields; every bush, every tree, almost every clod, for the larks were so many, seemed to have its songster. As for nightingales, I never knew so many in the most secluded country.

There are more round about London than in all the woodlands I used to ramble through. When people go into the country they really leave the birds behind them. It was the same, I found, after longer observation, with birds perhaps less widely known as with those universally recognised—such, for instance, as shrikes. The winter when the cry was raised that there were no birds, that the blackbirds and thrushes had left the lawns and must be dead, and how wicked it would be to take a nest next year, I had not the least difficulty in finding plenty of them.

They had simply gone to the water meadows, the brooks, and moist places generally. Every locality where running water kept the ground moist and permitted of movement among the creeping things which form these birds' food, was naturally resorted to. Thrushes and blackbirds, although they do not pack—that is, regularly fly in flocks—undoubtedly migrate when pressed by weather.

They are well known to arrive on the east coast from Norway in numbers as the cold increases. I see no reason why we may not suppose that in very severe and continued frost the thrushes and blackbirds round London fly westwards towards the milder side of the island. It seems to me that when, some years since, I used to stroll round the water meadows in a western county for snipes in frosty weather, the hedges were full of thrushes and blackbirds—quite full of them.

Now, though there were thrushes and blackbirds about the brooks by London last winter, there were few in the hedges generally. Had they, then, flown westwards? It is my belief that they had. They had left the hard-bound ground about London for the softer and moister lands farther west. They had crossed the rain-line. When frost prevents access to food in the east, thrushes and blackbirds move westwards, just as the fieldfares and redwings do.

That the fieldfares and redwings do so I can say with confidence, because, as they move in large flocks, there is no difficulty in tracing the direction in which they are going. They all went west when the severe weather began. On the southern side of London, at least in the districts I am best acquainted with, there was hardly a fieldfare or redwing to be seen for weeks and even months. Towards spring they came back, flying east for Norway. As thrushes and blackbirds move singly, and not with concerted action, their motions cannot be determined with such precision, but all the facts are in favour of the belief that they also went west.

That they were killed by the frost and snow I utterly refuse to credit. Some few, no doubt, were—I saw some greatly enfeebled by starvation— but not the mass. If so many had been destroyed their bodies must have been seen when there was no foliage to hide them, and no insects to quickly play the scavenger as in summer. Some were killed by cats; a few perhaps by rats, for in sharp winters they go down into the ditches, and I saw a dead redwing, torn and disfigured, at the mouth of a drain during the snow, where it might have been fastened on by a rat. But it is quite improbable that thousands died as was supposed.

Thrushes and blackbirds are not like rooks. Rooks are so bound by tradition and habit that they very rarely quit the locality where they were reared. Their whole lives are spent in the neighbourhood of the nest, trees, and the woods where they sleep. They may travel miles during the day, but they always come back to roost. These are the birds that suffer the most during long frosts and snows. Unable to break the chain that binds them to one spot, they die rather than desert it. A miserable time, indeed, they had of it that winter, but I never heard that any one proposed feeding the rooks, the very birds that wanted it most.

Swallows, again, were declared by many to be fewer. It is not at all unlikely that they were fewer. The wet season was unfavourable to them; still a good deal of the supposed absence of swallows may be through the observer not looking for them in the right place. If not wheeling in the sky, look for them over the water, the river, or great ponds; if not there, look along the moist fields or shady woodland meadows. They vary their haunts with the state of the atmosphere, which causes insects to be more numerous in one place at one time, and presently in another.

A very wet season is more fatal than the sharpest frost; it acts by practically reducing the births, leaving the ordinary death-rate to continue. Consequently, as the old birds die, there are none (or fewer) to supply their places. Once more let me express the opinion that there are as many small birds round London as in the country, and no measure is needed to protect the species at large. Protection, if needed, is required for the individual. Sweep the roads and lanes clear of the birdcatchers, but do not prevent a boy from taking a nest in the open fields or commons. If it were made illegal to sell full-grown birds, half the evil would be stopped at once if the law were enforced. The question is full of difficulties. To prevent or attempt to prevent the owner of a garden from shooting the bullfinches or blackbirds and so on that steal his fruit, or destroy his buds, is absurd. It is equally absurd to fine—what twaddle!—a lad for taking a bird's egg. The only point upon which I am fully clear is that the birdcatcher who takes birds on land not his own or in his occupation, on public property, as roads, wastes, commons, and so forth, ought to be rigidly put down. But as for the small birds as a mass, I am convinced that they will never cease out of the land.

It is not easy to progress far along this road, because every bird suggests so many reflections and recollections. Upon approaching the rising ground at Ewell green plovers or peewits become plentiful in the cornfields. In spring and early summer the flocks break up to some extent, and the scattered parties conduct their nesting operations in the pastures or on the downs. In autumn they collect together again, and flocks of fifty or more are commonly seen. Now and then a much larger flock comes down into the plain, wheeling to and fro, and presently descending upon an arable field, where they cover the ground.

NIGHTINGALE ROAD

The wayside is open to all, and that which it affords may be enjoyed without fee; therefore it is that I return to it so often. It is a fact that common hedgerows often yield more of general interest than the innermost recesses of carefully guarded preserves, which by day are frequently still, silent, and denuded of everything, even of game; nor can flowers flourish in such thick shade, nor where fir-needles cover the ground.

By the same wayside of which I have already spoken there is a birch copse, through which runs a road open to foot passengers, but not to wheel traffic, and also a second footpath. From these a little observation will show that almost all the life and interest of the copse is at, or near, the edge, and can be readily seen without trespassing a single yard. Sometimes, when it is quiet in the evening and the main highway is comparatively deserted, a hare comes stealing down the track through the copse, and after lingering there awhile crosses the highway into the stubble on the other side.

In one of these fields, just opposite the copse, a covey of partridges had their rendezvous, and I watched them from the road, evening after evening, issue one by one, calling as they appeared from a breadth of mangolds. Their sleeping-place seemed to be about a hundred yards from the wayside. Another arable field just opposite is bounded by the road with iron wire or railing, instead of a hedge, and the low mound in which the stakes are fixed swarmed one summer with ant-hills full of eggs, and a slight rustle in the corn as I approached told where the parent bird had just led her chicks from the feast to shelter.

Passing into the copse by the road, which is metalled but weed-grown from lack of use, the grasshoppers sing from the sward at the sides, but the birds are silent as the summer ends. Pink striped bells of convolvulus flower over the flints and gravel, the stones nearly hidden by their runners and leaves; yellow toadflax or eggs and bacon grew here till a weeding took place, since which it has not reappeared, but in its place viper's bugloss sprang up, a plant which was not previously to be found there. Hawkweeds, some wild vetches, white yarrow, thistles, and burdocks conceal the flints yet further, so that the track has the appearance of a green drive.

The slender birch and ash poles are hung with woodbine and wild hops, both growing in profusion. A cream-coloured wall of woodbine in flower extends in one spot, in another festoons of hops hang gracefully, and so thick as to hide everything beyond them. There is scarce a stole without its woodbine or hops; many of the poles, though larger than the arm, are scored with spiral grooves left by the bines. Under these bushes of woodbine the nightingales when they first arrive in spring are fond of searching for food, and dart on a grub with a low satisfied "kurr."

The place is so favourite a resort with these birds that it might well be called Nightingale Copse. Four or five may be heard singing at once on a warm May morning, and at least two may often be seen as well as heard at the same time. They sometimes sing from the trees, as well as from the bushes; one was singing one morning on an elm branch which projected over the road, and under which the van drivers jogged indifferently along. Sometimes they sing from the dark foliage of the Scotch firs.

As the summer wanes they haunt the hawthorn hedge by the roadside, leaving the interior of the copse, and may often be seen on the dry and dusty sward. When chiffchaff and willow-wren first come they remain in the treetops, but in the summer descend into the lower bushes, and, like the nightingales, come out upon the sward by the wayside. Nightingale Copse is also a great favourite with cuckoos. There are a few oaks in it, and in the meadows in the rear many detached hawthorn bushes, and two or three small groups of trees, chestnuts, lime, and elm. From the hawthorns to the elms, and from the elms to the oaks, the cuckoos continually circulate, calling as they fly.

One morning in May, while resting on a rail in the copse, I heard four calling close by, the furthest not a hundred yards distant, and as they continually changed their positions flying round there was always one in sight. They circled round, singing; the instant one ceased another took it up, a perfect madrigal. In the evening, at eight o'clock, I found them there again, still singing. The same detached groups of trees are much frequented by wood-pigeons, especially towards autumn.

Rooks prefer to perch on the highest branches, wood-pigeons more in the body of the tree, and when the boughs are bare of leaves a flock of the latter may be recognised in this way as far as the eye can see, and when the difference of colour is rendered imperceptible by distance. The wood-pigeon when perched has a rounded appearance; the rook a longer and sharper outline.

By one corner of the copse there is an oak, hollow within, but still green and flourishing. The hollow is black and charred; some mischievous boys must have lighted a fire inside it, just as the ploughboys do in the far away country. A little pond in the meadow close by is so overhung by another oak, and so surrounded with bramble and hawthorn, that the water lies in perpetual shade. It is just the spot where, if rabbits were about, one might be found sitting out on the bank under the brambles. This overhanging oak was broken by the famous October snow, 1880, further splintered by the gales of the next year, and its trunk is now split from top to bottom as if with wedges.

These meadows in spring are full of cowslips, and in one part the meadow-orchis flourishes. The method of making cowslip balls is universally known to children, from the most remote hamlet to the very verge of London, and the little children who dance along the green sward by the road here, if they chance to touch a nettle, at once search for a dock leaf to lay on it and assuage the smart. Country children, and indeed older folk, call the foliage of the knotted figwort cutfinger leaves, as they are believed to assist the cure of a cut or sore.

Raspberry suckers shoot up in one part of the copse; the fruit is doubtless eaten by the birds. Troops of them come here, travelling along the great hedge by the wayside, and all seem to prefer the outside trees and bushes to the interior of the copse. This great hedge is as wide as a country double mound, though it has but one ditch; the thick hawthorn, blackthorn, elder, and bramble—the oaks, elms, ashes, and firs form, in fact, almost a cover of themselves.

In the early spring, when the east wind rushes with bitter energy across the plains, this immense hedge, as far as it extends, shelters the wayfarer, the road being on the southern side, so that he can enjoy such gleams of sunshine as appear. In summer the place is, of course for the same reason, extremely warm, unless the breeze chances to come up strong from the west, when it sweeps over the open cornfields fresh and sweet. Stoats and weasels are common on the mound, or crossing the road to the corn; they seem more numerous in autumn, and I fear leveret and partridge are thinned by them.

Mice abound; in spring they are sometimes up in the blackthorn bushes, perhaps for the young buds. In summer they may often be heard rushing along the furrows across the wayside sward, scarce concealed by the wiry grass. Flowers are very local in habit; the spurge, for instance, which is common in a road parallel to this, is not to be seen, and not very much cow-parsnip, or "gix," one of the most freely-growing hedge plants, which almost chokes the mounds near by. Willowherbs, however, fill every place in the ditch here where they can find room between the bushes, and the arum is equally common, but the lesser celandine absent.

Towards evening, as the clover and vetches closed their leaves under the dew, giving the fields a different aspect and another green, I used occasionally to watch from here a pair of herons, sailing over in their calm serene way. Their flight was in the direction of the Thames, and they then passed evening after evening, but the following summer they did not come. One evening, later on in autumn, two birds appeared descending across the cornfields towards a secluded hollow where there was water, and, although at a considerable distance, from their manner of flight I could have no doubt they were teal.

The spotted leaves of the arum appeared in the ditches in this locality very nearly simultaneously with the first whistling of the blackbirds in February; last spring the chiffchaff sang soon after the flowering of the lesser celandine (not in this hedge, but near by), and the first swift was noticed within a day or two of the opening of the May bloom. Although not exactly, yet in a measure, the movements of plant and bird life correspond.

In a closely cropped hedge opposite this great mound (cropped because enclosing a cornfield) there grows a solitary shrub of the wayfaring tree. Though well known elsewhere, there is not, so far as I am aware, another bush of it for miles, and I should not have noticed this had not this part of the highway been so pleasant a place to stroll to and fro in almost all the year. The twigs of the wayfaring tree are covered with a mealy substance which comes off on the fingers when touched. A stray shrub or plant like this sometimes seems of more interest than a whole group.

For instance, most of the cottage gardens have foxgloves in them, but I had not observed any wild, till one afternoon near some woods I found a tall and beautiful foxglove, richer in colour than the garden specimens, and with bells more thickly crowded, lifting its spike of purple above the low cropped hawthorn. In districts where the soil is favourable to the foxglove it would not have been noticed, but here, alone and unexpected, it was welcomed. The bees in spring come to the broad wayside sward by the great mound to the bright dandelions; presently to the white clover, and later to the heaths.

There are about sixty wild flowers which grow freely along this road, namely, yellow agrimony, amphibious persicaria, arum, avens, bindweed, bird's foot lotus, bittersweet, blackberry, black and white bryony, brooklime, burdock, buttercups, wild camomile, wild carrot, celandine (the great and lesser), cinquefoil, cleavers, corn buttercup, corn mint, corn sowthistle, and spurrey, cowslip, cow-parsnip, wild parsley, daisy, dandelion, dead nettle, and white dog rose, and trailing rose, violets (the sweet and the scentless), figwort, veronica, ground ivy, willowherb (two sorts), herb Robert, honeysuckle, lady's smock, purple loosestrife, mallow, meadow-orchis, meadow-sweet, yarrow, moon daisy, St. John's wort, pimpernel, water plantain, poppy, rattles, scabious, self-heal, silverweed, sowthistle, stitchwort, teazles, tormentil, vetches, and yellow vetch.

To these may be added an occasional bacon and eggs, a few harebells (plenty on higher ground), the yellow iris, by the adjoining brook, and flowering shrubs and trees, as dogwood, gorse, privet, blackthorn, hawthorn, horse chestnut, besides wild hops, the horsetails on the mounds, and such plants as grow everywhere, as chickweed, groundsel, and so forth. A solitary shrub of mugwort grows at some distance, but in the same district, and in one hedgerow the wild guelder rose flourishes. Anemones and primroses are not found along or near this road, nor woodruff. At the first glance a list like this reads as if flowers abounded, but the reverse is the impression to those who frequent the place.

It is really a very short list, and as of course all of these do not appear at once there really is rather a scarcity of wild flowers, so far at least as variety goes. Just in the spring there is a burst of colour, and again in the autumn; but for the rest, if we set aside the roses in June, there seems quite an absence of flowers during the summer. The wayside is green, the ditches are green, the mounds green; if you enter and stroll round the meadows, they are green too, or white in places with umbelliferous plants, principally parsley and cow-parsnip. But these become monotonous. Therefore, I am constrained to describe it as a district somewhat lacking flowers, meaning, of course, in point of variety.

Compared with the hedges and fields of Wiltshire, Gloucestershire, Berkshire, and similar south-western localities, it seems flowerless. On the other hand, southern London can boast stretches of heath, which, when in full bloom, rival Scotch hillsides. These remarks are written entirely from a non-scientific point of view. Professional botanists may produce lists of thrice the length, and prove that all the flowers of England are to be found near London. But it will not alter the fact that to the ordinary eye the roads and lanes just south of London are in the middle of the summer comparatively bare of colour. They should be visited in spring and autumn.

Nor do the meadows seem to produce so many varieties of grass as farther to the south-west. But beetles of every kind and size, from the great stag beetle, helplessly floundering through the evening air and clinging to your coat, down to the green, bronze, and gilded species that hasten across the path, appear extremely numerous. Warm, dry sands, light soils, and furze and heath are probably favourable to them.

From this roadside I have seldom heard the corncrake, and never once the grasshopper lark. These two birds are so characteristic of the meadows in southwestern counties that a summer evening seems silent to me without the "crake, crake!" of the one and the singular sibilous rattle of the other. But they come to other places not far distant from the road, and one summer a grasshopper-lark could be heard in some meadows where I had not heard it the two preceding seasons. On the mounds field crickets cry persistently.

At the end of the hedge which is near a brook, a sedge-reedling takes up his residence in the spring. The sedge-reedlings here begin to call very early; the first date I have down is the 16th of April, which is, I think, some weeks before they begin in other localities. In one ditch beside the road (not in this particular hedge) there grows a fine bunch of reeds. Though watery, on account of the artificial drains from the arable fields, the spot is on much higher ground than the brook, and it is a little singular that while reeds flourish in this place they are not to be found by the brook.

The elms of the neighbourhood, wherever they can be utilised as posts, are unmercifully wired, wires twisted round, holes bored and the ends of wire driven in or staples inserted, and the same with the young oaks. Many trees are much disfigured from this cause, the bark is worn

47

off on many; and others, which have recovered, have bulging rings, where it swelled up over the iron. The heads of large nails and staples are easily discovered where the wire has disappeared, sometimes three or four, one above the other, in the same tree. A fine avenue of elms which shades part of a suburb appears to be dying by degrees—the too common fate of elms in such places.

How many beautiful trees have thus perished near London?—witness the large elms that once stood in Jews' Walk, at Sydenham. Barking the trunks for sheer wanton mischief is undoubtedly the cause in some cases, and it has been suggested that quicksilver has occasionally been inserted in gimlet holes. The mercury is supposed to work up the channels of the sap, and to prevent its flow.

But may not the ordinary conditions of suburban improvement often account for the decay of such trees without occult causes? Sewers carry away the water that used to moisten the roots, and being at some depth, they not only take the surface water of a storm before it has had time to penetrate, but drain the lower stratum completely. Then, gas-pipes frequently leak, so much so that the soil for yards is saturated and emits a smell of gas. Roots passing through such a soil can scarcely be healthy, and very probably, in making excavations for laying pipes the roots are cut through. The young trees that have been planted in some places are, I notice, often bored by grubs to an extraordinary extent, and will never make sound timber.

One July day, while walking on this road, I happened to look over a gateway and saw that a large and prominent mansion on the summit of some elevated ground had apparently disappeared. The day was very clear and bright, sunny and hot, and there was no natural vapour. But on the light north-east wind there came slowly towards me a bluish-yellow mist, the edge of which was clearly defined, and which blotted out distant objects and blurred those nearer at hand. The appearance of the open arable field over which I was looking changed as it approached.

In front of the wall of mist the sunshine lit the field up brightly, behind the ground was dull, and yet not in shadow. It came so slowly that its movement could be easily watched. When it went over me there was a perceptible coolness and a faint smell of damp smoke, and immediately the road, which had been white under the sunshine, took a dim, yellowish hue. The sun was not shut out nor even obscured, but the rays had to pass through a thicker medium. This haze was not thick enough to be called fog, nor was it the summer haze that in the country adds to the beauty of distant hills and woods.

It was clearly the atmosphere—not the fog—but simply the atmosphere of London brought out over the fields by a change in the wind, and prevented from diffusing itself by conditions of which nothing seems known. For at ordinary times the atmosphere of London diffuses itself in aerial space and is lost, but on this hot July day it came bodily and undiluted out into the cornfields. From its appearance I should say it would travel many miles in the same condition. In November fog seems seasonable: in hot and dry July this phenomenon was striking.

Along the road flocks of sheep continue to travel, some weary enough, and these, gravitating to the rear of the flock by reason of infirmity, lie down in the dust to rest, while their companions feed on the wayside sward. But the shepherds are careful of them, and do not hasten. Shepherds here often carry the pastoral crook. In districts far from the metropolis you may wander about for days, and with sheep all round you, never see a shepherd with a crook; but near town the pastoral staff is common.

These flocks appear to be on their way to the southern down farms, and, as I said before, the shepherds are tender over their sheep and careful not to press them. I regret that I cannot say the same about the bullocks, droves of which continually go by, often black cattle, and occasionally even the little Highland animals. The appearance of some of these droves is quite sufficient to indicate the treatment they have undergone. Staring eyes, heads continually turned from side to side, starting at everything, sometimes bare places on the shoulders, all tell the same tale of blows and brutal treatment.

Suburban streets which a minute before were crowded with ladies and children (most gentlemen are in town at midday) are suddenly vacated when the word passes that cattle are coming. People rush everywhere, into gardens, shops, back lanes, anywhere, as if the ringing scabbards of charging cavalry were heard, or the peculiar thumping rattle of rifles as they come to the "present" before a storm of bullets. It is no wonder that townsfolk exhibit a fear of cattle which makes their friends laugh when they visit the country after such experiences as these. This should be put down with a firm hand.

By the roadside here the hay tyers, who cut up the hayricks into trusses, use balances—a trifling matter, but sufficient to mark a difference, for in the west such men use a steelyard slung on a prong, the handle of the prong on the shoulder and the points stuck in the rick, with which to weigh the trusses. Wooden cottages, wooden barns, wooden mills are also characteristic.

Mouchers come along the road at all times and seasons, gathering sacksful of dandelions in spring, digging up fern roots and cowslip mars for sale, cutting briars for standard roses, gathering water-cresses and mushrooms, and in the winter cutting rushes.

There is a rook with white feathers in the wing which belongs to an adjacent rookery, and I have observed a blackbird also streaked with white. One January day, when the snow was on the ground and the frost was sharp, when the pale sun seemed to shine brightest round the rim of the disk, as if there were a band of stronger light there, I saw a white animal under a heap of poles by the wayside, near the great hedge I have mentioned. It immediately concealed itself, but, thinking that it was a ferret gone astray, I waited, and presently the head and neck were cautiously protruded.

I made the usual call with the lips, but the creature instantly returned to cover. I waited again, hiding this time, and after an interval the creature moved and hastened away from the poles, where it was, in a measure, exposed, to the more secure shelter of some bushes. Then I saw that it was of a clear white, while so-called white ferrets are usually a dingy yellow, and the white tail was tipped with black. From these circumstances, and from the timidity and anxious desire to escape observation, I could only conclude that it was a white stoat.

Stoats, as remarked previously, are numerous in these hedges, and it was quite possible for a white one to be among them. The white stoat may be said to exactly resemble the ermine. The interest of the circumstance arises not from its rarity, but from its occurring so near the metropolis.

A BROOK

Some low wooden rails guarding the approach to a bridge over a brook one day induced me to rest under an aspen, with my back against the tree. Some horse-chestnuts, beeches, and alders grew there, fringing the end of a long plantation of willow stoles which extended in the rear following the stream. In front, southwards, there were open meadows and cornfields, over which shadow and sunshine glided in succession as the sweet westerly wind carried the white clouds before it.

The brimming brook, as it wound towards me through the meads, seemed to tremble on the verge of overflowing, as the crown of wine in a glass rises yet does not spill. Level with the green grass, the water gleamed as though polished where it flowed smoothly, crossed with the dark shadows of willows which leaned over it. By the bridge, where the breeze rushed through the arches, a ripple flashed back the golden rays. The surface by the shore slipped towards a side hatch and passed over in a liquid curve, clear and unvarying, as if of solid crystal, till shattered on the stones, where the air caught up and played with the sound of the bubbles as they broke.

Beyond the green slope of corn, a thin, soft vapour hung on the distant woods, and hid the hills. The pale young leaves of the aspen rustled faintly, not yet with their full sound; the sprays of the horse-chestnut, drooping with the late frosts, could not yet keep out the sunshine with their broad green. A white spot on the footpath yonder was where the bloom had fallen from a blackthorn bush.

The note of the tree-pipit came from over the corn—there were some detached oaks away in the midst of the field, and the birds were doubtless flying continually up and down between the wheat and the branches. A willow-wren sang plaintively in the plantation behind, and once a cuckoo called at a distance. How beautiful is the sunshine! The very dust of the road at my feet seemed to glow with whiteness, to be lit up by it, and to become another thing. This spot henceforward was a place of pilgrimage.

Looking that morning over the parapet of the bridge, down stream, there was a dead branch at the mouth of the arch, it had caught and got fixed while it floated along. A quantity of aquatic weeds coming down the stream had drifted against the branch and remained entangled in it. Fresh weeds were still coming and adding to the mass, which had attracted a water-rat.

Perched on the branch the little brown creature bent forward over the surface, and with its two forepaws drew towards it the slender thread of a weed, exactly as with hands. Holding the thread in the paws, it nibbled it, eating the sweet and tender portion, feeding without fear though but a few feet away, and precisely beneath me.

In a minute the surface of the current was disturbed by larger ripples. There had been a ripple caused by the draught through the arch, but this was now increased. Directly afterwards a moorhen swam out, and began to search among the edge of the tangled weeds. So long as I was perfectly still the bird took no heed, but at a slight movement instantly scuttled back under the arch. The water-rat, less timorous, paused, looked round, and returned to feeding.

Crossing to the other side of the bridge, up stream, and looking over, the current had scooped away the sand of the bottom by the central pier, exposing the brickwork to some depth—the same undermining process that goes on by the piers of bridges over great rivers. Nearer the shore the sand has silted up, leaving it shallow, where water-parsnip and other weeds joined, as it were, the verge of the grass and the stream. The sunshine reflected from the ripples on this, the southern side, continually ran with a swift, trembling motion up the arch.

53

Penetrating the clear water, the light revealed the tiniest stone at the bottom: but there was no fish, no water-rat, or moorhen on this side. Neither on that nor many succeeding mornings could anything be seen there; the tail of the arch was evidently the favourite spot. Carefully looking over that side again, the moorhen who had been out rushed back; the water-rat was gone. Were there any fish? In the shadow the water was difficult to see through, and the brown scum of spring that lined the bottom rendered everything uncertain.

By gazing steadily at a stone my eyes presently became accustomed to the peculiar light, the pupils adjusted themselves to it, and the brown tints became more distinctly defined. Then sweeping by degrees from a stone to another, and from thence to a rotting stick embedded in the sand, I searched the bottom inch by inch. If you look, as it were at large—at everything at once—you see nothing. If you take some object as a fixed point, gaze all around it, and then move to another, nothing can escape.

Even the deepest, darkest water (not, of course, muddy) yields after a while to the eye. Half close the eyelids, and while gazing into it let your intelligence rather wait upon the corners of the eye than on the glance you cast straight forward. For some reason when thus gazing the edge of the eye becomes exceedingly sensitive, and you are conscious of slight motions or of a thickness—not a defined object, but a thickness which indicates an object—which is otherwise quite invisible.

The slow feeling sway of a fish's tail, the edges of which curl over and grasp the water, may in this manner be identified without being positively seen, and the dark outline of its body known to exist against the equally dark water or bank. Shift, too, your position according to the fall of the light, just as in looking at a painting. From one point of view the canvas shows little but the presence of paint and blurred colour, from another at the side the picture stands out.

Sometimes the water can be seen into best from above, sometimes by lying on the sward, now by standing back a little way, or crossing to the opposite shore. A spot where the sunshine sparkles with dazzling gleam is perhaps perfectly inpenetrable till you get the other side of the ripple, when the same rays that just now baffled the glance light up the bottom as if thrown from a mirror for the purpose. I convinced myself that there was nothing here, nothing visible at present—not so much as a stickleback.

Yet the stream ran clear and sweet, and deep in places. It was too broad for leaping over. Down the current sedges grew thickly at a curve: up the stream the young flags were rising; it had an inhabited look, if such a term may be used, and moorhens and water-rats were about but no fish. A wide furrow came along the meadow and joined the stream from the side. Into this furrow, at flood time, the stream overflowed farther up, and irrigated the level sward.

At present it was dry, its course, traced by the yellowish and white hue of the grasses in it only recently under water, contrasting with the brilliant green of the sweet turf around. There was a marsh marigold in it, with stems a quarter of an inch thick; and in the grass on the verge, but just beyond where the flood reached, grew the lilac-tinted cuckoo flowers, or cardamine.

The side hatch supplied a pond, which was only divided from the brook by a strip of sward not more than twenty yards across. The surface of the pond was dotted with patches of scum that had risen from the bottom. Part at least of it was shallow, for a dead branch blown from an elm projected above the water, and to it came a sedge-reedling for a moment. The sedge-reedling is so fond of sedges, and reeds, and thick undergrowth, that though you hear it perpetually within a few yards it is not easy to see one. On this bare branch the bird was well displayed, and the streak by the eye was visible; but he stayed there for a second or two only, and then back again to the sedges and willows.

There were fish I felt sure as I left the spot and returned along the dusty road, but where were they?

On the sward by the wayside, among the nettles and under the bushes, and on the mound the dark green arum leaves grew everywhere, sometimes in bunches close together. These bunches varied—in one place the leaves were all spotted with black irregular blotches; in another the leaves were without such markings. When the root leaves of the arum first push up they are closely rolled together in a pointed spike.

This, rising among the dead and matted leaves of the autumn, occasionally passes through holes in them. As the spike grows it lifts the dead leaves with it, which hold it like a ring and prevent it from unfolding. The force of growth is not sufficiently strong to burst the bond asunder till the green leaves have attained considerable size.

A little earlier in the year the chattering of magpies would have been heard while looking for the signs of spring, but they were now occupied with their nests. There are several within a short distance, easily distinguished in winter, but somewhat hidden now by the young leaves. Just before they settled down to housekeeping there was a great chattering and fluttering and excitement, as they chased each other from elm to elm.

Four or five were then often in the same field, some in the trees, some on the ground, their white and black showing distinctly on the level brown earth recently harrowed or rolled. On such a surface birds are visible at a distance; but when the blades of the corn begin to reach any height such as alight are concealed. In many districts of the country that might be called wild and lonely, the magpie is almost extinct. Once now and then a pair may be observed, and those who know their haunts can, of course, find them, but to a visitor passing through, there seems none. But here, so near the metropolis, the magpies are common, and during an hour's walk their cry is almost sure to be heard. They have, however, their favourite locality, where they are much more frequently seen.

Coming to my seat under the aspen by the bridge week after week, the burdocks by the wayside gradually spread their leaves, and the procession of the flowers went on. The dandelion, the lesser celandine, the marsh marigold, the coltsfoot, all yellow, had already led the van, closely accompanied by the purple ground-ivy, the red dead-nettle, and the daisy; this last a late comer in the neighbourhood. The blackthorn, the horse-chestnut, and the hawthorn came, and the meadows were golden with the buttercups.

Once only had I noticed any indication of fish in the brook; it was on a warm Saturday afternoon, when there was a labourer a long way up the stream, stooping in a peculiar manner near the edge of the water with a stick in his hand. He was, I felt sure, trying to wire a spawning jack, but did not succeed. Many weeks had passed, and now there came (as the close time for coarse fish expired) a concourse of anglers to the almost stagnant pond fed by the side hatch.

Well-dressed lads with elegant and finished tackle rode up on their bicycles, with their rods slung at their backs. Hoisting the bicycles over the gate into the meadow, they left them leaning against the elms, fitted their rods and fished in the pond. Poorer boys, with long wands cut from the hedge and ruder lines, trudged up on foot, sat down on the sward and watched their corks by the hour together. Grown men of the artisan class, covered with the dust of many miles' tramping, came with their luncheons in a handkerchief, and set about their sport with a quiet earnestness which argued long if desultory practice.

In fine weather there were often a dozen youths and four or five men standing, sitting, or kneeling on the turf along the shore of the pond, all intent on their floats, and very nearly silent. People driving along the highway stopped their traps, and carts, and vans a minute or two to watch them: passengers on foot leaned over the gate, or sat down and waited expectantly.

Sometimes one of the more venturesome anglers would tuck up his trousers and walk into the shallow water, so as to be able to cast his bait under the opposite bank, where it was deep. Then an ancient and much battered punt was discovered aground in a field at some distance, and dragged to the pond. One end of the punt had quite rotted away, but by standing at the other, so as to depress it there and lift the open end above the surface, two, or even three, could make a shift to fish from it.

The silent and motionless eagerness with which these anglers dwelt upon their floats, grave as herons, could not have been exceeded. There they were day after day, always patient and always hopeful. Occasionally a small catch—a mere "bait "—was handed round for inspection; and once a cunning fisherman, acquainted with all the secrets of his craft, succeeded in drawing forth three perch, perhaps a quarter of a pound each, and one slender eel. These made quite a show, and were greatly admired; but I never saw the same man there again. He was satisfied.

As I sat on the white rail under the aspen, and inhaled the scent of the beans flowering hard by, there was a question which suggested itself to me, and the answer to which I never could supply. The crowd about the pond all stood with their backs to the beautiful flowing brook. They had before them the muddy banks of the stagnant pool, on whose surface patches of scum floated.

Behind them was the delicious stream, clear and limpid, bordered with sedge and willow and flags, and overhung with branches. The strip of sward between the two waters was certainly not more than twenty yards; there was no division hedge, or railing, and evidently no preservation, for the mouchers came and washed their water-cress which they had gathered in the ditches by the side hatch, and no one interfered with them.

There was no keeper or water bailiff, not even a notice board. Policemen, on foot and mounted, passed several times daily, and, like everybody else, paused to see the sport, but said not a word. Clearly, there was nothing whatever to prevent any of those present from angling in the stream; yet they one and all, without exception, fished in the pond. This seemed to me a very remarkable fact.

After a while I noticed another circumstance; nobody ever even looked into the stream or under the arches of the bridge. No one spared a moment from his float amid the scum of the pond, just to stroll twenty paces and glance at the swift current. It appeared from this that the pond had a reputation for fish, and the brook had not. Everybody who had angled in the pond recommended his friends to go and do likewise. There were fish in the pond.

So every fresh comer went and angled there, and accepted the fact that there were fish. Thus the pond obtained a traditionary reputation, which circulated from lip to lip round about. I need not enlarge on the analogy that exists in this respect between the pond and various other things.

By implication it was evidently as much understood and accepted on the other hand that there was nothing in the stream. Thus I reasoned it out, sitting under the aspen, and yet somehow the general opinion did not satisfy me. There must be something in so sweet a stream. The sedges by the shore, the flags in the shallow, slowly swaying from side to side with the current, the sedge-reedlings calling, the moorhens and water-rats, all gave an air of habitation.

One morning, looking very gently over the parapet of the bridge (down stream) into the shadowy depth beneath, just as my eyes began to see the bottom, something like a short thick dark stick drifted out from the arch, somewhat sideways. Instead of proceeding with the current, it had hardly cleared the arch when it took a position parallel to the flowing water and brought up. It was thickest at the end that faced the stream; at the other there was a slight motion as if caused by the current against a flexible membrane, as it sways a flag. Gazing down intently into the shadow the colour of the sides of the fish appeared at first not exactly uniform, and presently these indistinct differences resolved themselves into spots. It was a trout, perhaps a pound and a half in weight.

His position was at the side of the arch, out of the rush of the current, and almost behind the pier, but where he could see anything that came floating along under the culvert. Immediately above him but not over was the mass of weeds tangled in the dead branch. Thus in the shadow of the bridge and in the darkness under the weeds he might easily have escaped notice. He was, too, extremely wary. The slightest motion was enough to send him instantly under the arch; his cover was but a foot distant, and a trout shoots twelve inches in a fraction of time.

The summer advanced, the hay was carted, and the wheat ripened. Already here and there the reapers had cut portions of the more forward corn. As I sat from time to time under the aspen, within hearing of the murmuring water, the thought did rise occasionally that it was a pity to leave the trout there till some one blundered into the knowledge of his existence.

There were ways and means by which he could be withdrawn without any noise or publicity. But, then, what would be the pleasure of securing him, the fleeting pleasure of an hour, compared to the delight of seeing him almost day by day? I watched him for many weeks, taking great precautions that no one should observe how continually I looked over into the water there. Sometimes after a glance I stood with my back to the wall as if regarding an object on the other side. If any one was following me, or appeared likely to peer over the parapet, I carelessly struck the top of the wall with my stick in such a manner that it should project, an action sufficient to send the fish under the arch. Or I raised my hat as if heated, and swung it so that it should alarm him.

If the coast was clear when I had looked at him still I never left without sending him under the arch in order to increase his alertness. It was a relief to know that so many persons who went by wore tall hats, a safeguard against their seeing anything, for if they approached the shadow of the tall hat reached out beyond the shadow of the parapet, and was enough to alarm him before they could look over. So the summer passed, and, though never free from apprehensions, to my great pleasure without discovery.

A LONDON TROUT

The sword-flags are rusting at their edges, and their sharp points are turned. On the matted and entangled sedges lie the scattered leaves which every rush of the October wind hurries from the boughs. Some fall on the water and float slowly with the current, brown and yellow spots on the dark surface. The grey willows bend to the breeze; soon the osier beds will look reddish as the wands are stripped by the gusts. Alone the thick polled alders remain green, and in their shadow the brook is still darker. Through a poplar's thin branches the wind sounds as in the rigging of a ship; for the rest, it is silence.

The thrushes have not forgotten the frost of the morning, and will not sing at noon; the summer visitors have flown and the moorhens feed quietly. The plantation by the brook is silent, for the sedges, though they have drooped and become entangled, are not dry and sapless yet to rustle loudly. They will rustle dry enough next spring, when the sedge-birds come. A long withey-bed borders the brook and is more resorted to by sedge-reedlings, or sedge-birds, as they are variously called, than any place I know, even in the remotest country.

Generally it has been difficult to see them, because the withey is in leaf when they come, and the leaves and sheaves of innumerable rods hide them, while the ground beneath is covered by a thick growth of sedges and flags, to which the birds descend. It happened once, however, that the withey stoles had been polled, and in the spring the boughs were short and small. At the same time, the easterly winds checked the sedges, so that they were hardly half their height, and the flags were thin, and not much taller, when the sedge-birds came, so that they for once found but little cover, and could be seen to advantage.

There could not have been less than fifteen in the plantation, two frequented some bushes beside a pond near by, some stayed in scattered willows farther down the stream. They sang so much they scarcely seemed to have time to feed. While approaching one that was singing by gently walking on the sward by the roadside, or where thick dust deadened the footsteps, suddenly another would commence in the low thorn hedge on a branch, so near that it could be touched with a walking-stick. Yet though so near the bird was not wholly visible—he was partly concealed behind a fork of the bough. This is a habit of the sedge-birds. Not in the least timid, they chatter at your elbow, and yet always partially hidden.

If in the withey, they choose a spot where the rods cross or bunch together. If in the sedges, though so close it seems as if you could reach forward and catch him, he is behind the stalks. To place some obstruction between themselves and any one passing is their custom: but that spring, as the foliage was so thin, it only needed a little dexterity in peering to get a view. The sedge-bird perches aside, on a sloping willow rod, and, slightly raising his head, chatters, turning his bill from side to side. He is a very tiny bird, and his little eye looks out from under a yellowish streak. His song at first sounds nothing but chatter.

After listening a while the ear finds a scale in it—an arrangement and composition—so that, though still a chatter, it is a tasteful one. At intervals he intersperses a chirp, exactly the same as that of the sparrow, a chirp with a tang in it. Strike a piece of metal, and besides the noise of the blow, there is a second note, or tang. The sparrow's chirp has such a note sometimes, and the sedge-bird brings it in—tang, tang, tang. This sound has given him his country name of brook-sparrow, and it rather spoils his song. Often the moment he has concluded he starts for another willow stole, and as he flies begins to chatter when halfway across, and finishes on a fresh branch.

But long before this another bird has commenced to sing in a bush adjacent; a third takes it up in the thorn hedge; a fourth in the bushes across the pond; and from farther down the stream comes a faint and distant chatter. Ceaselessly the competing gossip goes on the entire day and most of the night; indeed, sometimes all night through. On a warm spring morning, when the sunshine pours upon the willows, and even the white dust of the road is brighter, bringing out the shadows in clear definition, their lively notes and quick motions make a pleasant commentary on the low sound of the stream rolling round the curve.

A moorhen's call comes from the hatch. Broad yellow petals of marsh-marigold stand up high among the sedges rising from the greyish-green ground, which is covered with a film of sun-dried aquatic grass left dry by the retiring waters. Here and there are lilac-tinted cuckoo-flowers, drawn up on taller stalks than those that grow in the meadows. The black flowers of the sedges are powdered with yellow pollen; and dark green sword-flags are beginning to spread their fans. But just across the road, on the topmost twigs of birch poles, swallows twitter in the tenderest tones to their loves. From the oaks in the meadows on that side titlarks mount above the highest bough and then descend, sing, sing, singing, to the grass.

A jay calls in a circular copse in the midst of the meadow; solitary rooks go over to their nests in the elms on the hill; cuckoos call, now this way and now that, as they travel round. While leaning on the grey and lichen-hung rails by the brook, the current glides by, and it is the motion of the water and its low murmur which renders the place so idle; the sunbeams brood, the air is still but full of song. Let us, too, stay and watch the petals fall one by one from a wild apple and float down on the stream.

But now in autumn the haws are red on the thorn, the swallows are few as they were in the earliest spring; the sedge-birds have flown, and the redwings will soon be here. The sharp points of the sword-flags are turned, their edges rusty, the forget-me-nots are gone. October's winds are too searching for us to linger beside the brook, but still it is pleasant to pass by and remember the summer days. For the year is never gone by; in a moment we can recall the sunshine we enjoyed in May, the roses we gathered in June, the first wheatear we plucked as the green corn filled. Other events go by and are forgotten, and even the details of our own lives, so immensely important to us at the moment, in time fade from the memory till the date we fancied we should never forget has to be sought in a diary. But the year is always with us; the months are familiar always; they have never gone by.

So with the red haws around and the rustling leaves it is easy to recall the flowers. The withey plantation here is full of flowers in summer; yellow iris flowers in June when midsummer comes, for the iris loves a thunder-shower. The flowering flag spreads like a fan from the root, the edges overlap near the ground, and the leaves are broad as sword-blades, indeed the plant is one of the largest that grows wild. It is quite different from the common flag with three grooves—bayonet shape—which appears in every brook. The yellow iris is much more local, and in many country streams may be sought for in vain, so that so fine a display as may be seen here seemed almost a discovery to me.

They were finest in the year of rain, 1879, that terrible year which is fresh in the memory of all who have any interest in out-of-door matters. At midsummer the plantation was aglow with iris bloom. The large yellow petals were everywhere high above the sedge; in one place a dozen, then two or three, then one by itself, then another bunch. The marsh was a foot deep in water, which could only be seen by parting the stalks of the sedges, for it was quite hidden under them. Sedges and flags grew so thick that everything was concealed except the yellow bloom above.

One bunch grew on a bank raised a few inches above the flood which the swollen brook had poured in, and there I walked among them; the leaves came nearly up to the shoulder, the golden flowers on the stalks stood equally high. It was a thicket of iris. Never before had they risen to such a height; it was like the vegetation of tropical swamps, so much was everything drawn up by the continual moisture. Who could have supposed that such a downpour as occurred that summer would have had the effect it had upon flowers? Most would have imagined that the excessive rain would have destroyed them; yet never was there such floral beauty as that year. Meadow-orchis, buttercups, the yellow iris, all the spring flowers came forth in extraordinary profusion. The hay was spoiled, the farmers ruined, but their fields were one broad expanse of flower.

As that spring was one of the wettest, so that of the year in present view was one of the driest, and hence the plantation between the lane and the brook was accessible, the sedges and flags short, and the sedge-birds visible. There is a beech in the plantation standing so near the verge of the stream that its boughs droop over. It has a number of twigs around the stem—as a rule the beech-bole is clear of boughs, but some which are of rather stunted growth are fringed with them. The leaves on the longer boughs above fall off and voyage down the brook, but those on the lesser twigs beneath, and only a little way from the ground, remain on, and rustle, dry and brown, all through the winter.

Under the shelter of these leaves, and close to the trunk, there grew a plant of flag—the tops of the flags almost reached to the leaves—and all the winter through, despite the frosts for which it was remarkable, despite the snow and the bitter winds which followed, this plant remained green and fresh. From this beech in the morning a shadow stretches to a bridge across the brook, and in that shadow my trout used to lie. The bank under the drooping boughs forms a tiny cliff a foot high, covered with moss, and here I once observed shrew mice diving and racing about. But only once, though I frequently passed the spot; it is curious that I did not see them afterwards.

Just below the shadow of the beech there is a sandy, oozy shore, where the footprints of moorhens are often traceable. Many of the trees of the plantation stand in water after heavy rain; their leaves drop into it in autumn, and, being away from the influence of the current, stay and soak, and lie several layers thick. Their edges overlap, red, brown, and pale yellow, with the clear water above and shadows athwart it, and dry white grass at the verge. A horse-chestnut drops its fruit in the dusty road; high above its leaves are tinted with scarlet.

It was at the tail of one of the arches of the bridge over the brook that my favourite trout used to lie. Sometimes the shadow of the beech came as far as his haunts, that was early in the morning, and for the rest of the day the bridge itself cast a shadow. The other parapet faces the south, and looking down from it the bottom of the brook is generally visible, because the light is so strong. At the bottom a green plant may be seen waving to and fro in summer as the current sways it. It is not a weed or flag, but a plant with pale green leaves, and looks as if it had come there by some chance; this is the water-parsnip.

By the shore on this, the sunny side of the bridge, a few forget-me-nots grow in their season, water crow's-foot flowers, flags lie along the surface and slowly swing from side to side like a boat at anchor. The breeze brings a ripple, and the sunlight sparkles on it; the light reflected dances up the piers of the bridge. Those that pass along the road are naturally drawn to this bright parapet where the brook winds brimming full through green meadows. You can see right to the bottom; you can see where the rush of the water has scooped out a deeper channel under the arches, but look as long as you like there are no fish.

The trout I watched so long, and with such pleasure, was always on the other side, at the tail of the arch, waiting for whatever might come through to him. There in perpetual shadow he lay in wait, a little at the side of the arch, scarcely ever varying his position except to dart a yard up under the bridge to seize anything he fancied, and drifting out again to bring up at his anchorage. If people looked over the parapet that side they did not see him; they could not see the bottom there for the shadow, or if the summer noonday cast a strong beam even then it seemed to cover the surface of the water with a film of light which could not be seen through. There are some aspects from which even a picture hung on the wall close at hand cannot be seen. So no one saw the trout; if any one more curious leant over the parapet he was gone in a moment under the arch.

Folk fished in the pond about the verge of which the sedge-birds chattered, and but a few yards distant; but they never looked under the arch on the northern and shadowy side, where the water flowed beside the beech. For three seasons this continued. For three summers I had the pleasure to see the trout day after day whenever I walked that way, and all that time, with fishermen close at hand, he escaped notice, though the place was not preserved. It is wonderful to think how difficult it is to see anything under one's very eyes, and thousands of people walked actually and physically right over the fish.

However, one morning in the third summer, I found a fisherman standing in the road and fishing over the parapet in the shadowy water. But he was fishing at the wrong arch, and only with paste for roach. While the man stood there fishing, along came two navvies; naturally enough they went quietly up to see what the fisherman was doing, and one instantly uttered an exclamation. He had seen the trout. The man who was fishing with paste had stood so still and patient that the trout, re-assured, had come out, and the navvy—trust a navvy to see anything of the kind—caught sight of him.

The navvy knew how to see through water. He told the fisherman, and there was a stir of excitement, a changing of hooks and bait. I could not stay to see the result, but went on, fearing the worst. But he did not succeed; next day the wary trout was there still, and the next, and the next. Either this particular fisherman was not able to come again, or was discouraged; at any rate, he did not try again. The fish escaped, doubtless more wary than ever.

In the spring of the next year the trout was still there, and up to the summer I used to go and glance at him. This was the fourth season, and still he was there; I took friends to look at this wonderful fish, which defied all the loafers and poachers, and above all, surrounded himself not only with the shadow of the bridge, but threw a mental shadow over the minds of passers-by, so that they never thought of the possibility of such a thing as trout. But one morning something happened. The brook was dammed up on the sunny side of the bridge, and the water let off by a side-hatch, that some accursed main or pipe or other horror might be laid across the bed of the stream somewhere far down.

Above the bridge there was a brimming broad brook, below it the flags lay on the mud, the weeds drooped, and the channel was dry. It was dry up to the beech tree. There, under the drooping boughs of the beech, was a small pool of muddy water, perhaps two yards long, and very narrow—a stagnant muddy pool, not more than three or four inches deep. In this I saw the trout. In the shallow water, his back came up to the surface (for his fins must have touched the mud sometimes)—once it came above the surface, and his spots showed as plain as if you had held him in your hand. He was swimming round to try and find out the reason of this sudden stinting of room.

Twice he heaved himself somewhat on his side over a dead branch that was at the bottom, and exhibited all his beauty to the air and sunshine. Then he went away into another part of the shallow and was hidden by the muddy water. Now under the arch of the bridge, his favourite arch, close by there was a deep pool, for, as already mentioned, the scour of the current scooped away the sand and made a hole there. When the stream was shut off by the dam above this hole remained partly full. Between this pool and the shallow under the beech there was sufficient connection for the fish to move into it.

My only hope was that he would do so, and as some showers fell, temporarily increasing the depth of the narrow canal between the two pools, there seemed every reason to believe that he had got to that under the arch. If now only that accursed pipe or main, or whatever repair it was, could only be finished quickly, even now the trout might escape! Every day my anxiety increased, for the intelligence would soon get about that the brook was dammed up, and any pools left in it would be sure to attract attention.

Sunday came, and directly the bells had done ringing four men attacked the pool under the arch. They took off shoes and stockings and waded in, two at each end of the arch. Stuck in the mud close by was an eel-spear. They churned up the mud, wading in, and thickened and darkened it as they groped under. No one could watch these barbarians longer.

Is it possible that he could have escaped? He was a wonderful fish, wary and quick. Is it just possible that they may not even have known that a trout was there at all; but have merely hoped for perch, or tench, or eels? The pool was deep and the fish quick—they did not bale it, might he have escaped? Might they even, if they did find him, have mercifully taken him and placed him alive in some other water nearer their homes? Is it possible that he may have almost miraculously made his way down the stream into other pools?

There was very heavy rain one night, which might have given him such a chance. These "mights," and "ifs," and "is it possible" even now keep alive some little hope that some day I may yet see him again. But that was in the early summer. It is now winter, and the beech has brown spots. Among the limes the sedges are matted and entangled, the sword-flags rusty; the rooks are at the acorns, and the plough is at work in the stubble. I have never seen him since. I never failed to glance over the parapet into the shadowy water. Somehow it seemed to look colder, darker, less pleasant than it used to do. The spot was empty, and the shrill winds whistled through the poplars.

A BARN

A broad red roof of tile is a conspicuous object on the same road which winds and turns in true crooked country fashion, with hedgerows, trees, and fields on both sides, and scarcely a dwelling visible. It is not, indeed, so crooked as a lane in Gloucestershire, which I verily believe passes the same tree thrice, but the curves are frequent enough to vary the view pleasantly.

Approaching from either direction, on turning a certain corner a great red roof rises high above the hedges, and the line of its ridge is seen every way through the trees. With this old barn, as with so much of the architecture of former times, the roof is the most important part. The gables, for instance, of Elizabethan houses occupy the eye far more than the walls; and so, too, with the antique halls that still exist. The roof of this old barn is itself the building; the roof and the doors, for the sweeping slope of the tiles comes down within reach of the hand, while the great doors extend half-way to the ridge.

By the low black wooden walls a little chaff has been spilt, and has blown out and mingles with the dust of the road. Loose straws lie across the footpath, trodden flat by passing feet; straws have wandered across the road and lodged on the mound, and others have roamed still farther round the corner. Between the gatepost and the wall that encloses the rickyard more straws are jammed, and yet more are borne up by the nettles beneath it.

Mosses have grown over the old red brick wall, both on the top and following the lines of the mortar, and bunches of wall grasses flourish along the top. The wheat, and barley, and hay carted home to the rickyard contain the seeds of innumerable plants, many of which, dropping to the ground, come up next year. The trodden earth round where the ricks stood seems favourable to their early appearance; the first poppy blooms here, though its colour is paler than those which come afterwards in the fields.

In spring most of the ricks are gone, threshed and sold, but there remains the vast pile of straw—always straw—and the three-cornered stump of a hay-rick which displays bands of different hues, one above the other, like the strata of a geological map. Some of the hay was put up damp, some in good condition, and some had been browned by bad weather before being carted.

About the straw-rick, and over the chaff that everywhere strews the earth, numerous fowls search, and by the gateway Chanticleer proudly stands, tall and upright, the king of the rickyard still, as he and his ancestors have been these hundreds of years. Under the granary, which is built on stone staddles, to exclude the mice, some turkeys are huddled together calling occasionally for a "halter," and beyond them the green, glossy neck of a drake glistens in the sunshine.

When the corn is high, and sometimes before it is well up, the doors of the barn are daily open, and shock-headed children peer over the hatch. There are others within playing and tumbling on a heap of straw—always straw—which is their bed at night. The sacks which form their counterpane are rolled aside, and they have half the barn for their nursery. If it is wet, at least one great girl and the mother will be there too, gravely sewing, and sitting where they can see all that goes along the road.

A hundred yards away, in a corner of an arable field, the very windiest and most draughty that could be chosen, where the hedge is cut down so that it can barely be called a hedge, and where the elms draw the wind, the men of the family crowd over a smoky fire. In the wind and rain the fire could not burn at all had they not by means of a stick propped up a hurdle to windward, and thus sheltered it. As it is there seems no flame, only white embers and a flow of smoke, into which the men from time to time cast the dead wood they have gathered. Here the pot is boiled and the cooking accomplished at a safe distance from the litter and straw of the rickyard.

These people are Irish, who come year after year to the same barn for the hoeing and the harvest, travelling from the distant West to gather agricultural wages on the verge of the metropolis.

In fine summer weather, beside the usual business traffic, there goes past this windy bare corner a constant stream of pleasure-seekers, heavily laden four-in-hands, tandems, dog-carts, equestrians, and open carriages, filled with well-dressed ladies. They represent the abundant gold of trade and commerce. In their careless luxury they do not notice—how should they?—the smoky fire in the barren corner, or the shock-headed children staring at the equipages over the hatch at the barn.

Within a mile there is a similar fire, which by day is not noticeable, because the spot is under a hedge two meadows back from the road. At night it shows brightly, and even as late as eleven o'clock dusky figures may be seen about it, as if the family slept in the open air. A third fire is kept up in the same neighbourhood, but in a different direction, in a meadow bordering on a lonely lane. There is a thatched shed behind the hedge, which is the sleeping-place—the fire burns some forty yards away. Still another shines at night in an open arable field, where is a barn.

One day I observed a farmer's courtyard completely filled with groups of men, women, and children, who had come travelling round to do the harvesting. They had with them a small cart or van—not of the kind which the show folk use as movable dwellings, but for the purpose of carrying their pots, pans, and the like. The greater number carry their burdens on their backs, trudging afoot.

A gang of ten or twelve once gathered round me to inquire the direction of some spot they desired to reach. A powerful-looking woman, with reaping-hook in her hand and cooking implements over her shoulder, was the speaker. The rest did not appear to know a word of English, and her pronunciation was so peculiar that it was impossible to understand what she meant except by her gestures. I suppose she wanted to find a farm, the name of which I could not get at, and then perceiving she was not understood her broad face flushed red and she poured out a flood of Irish in her excitement. The others chimed in, and the din redoubled. At last I caught the name of a town and was thus able to point the way.

About harvest time it is common to meet an Irish labourer dressed in the national costume: a tall, upright fellow with a long-tailed coat, breeches, and worsted stockings. He walks as upright as if drilled, with a quick easy gait and springy step, quite distinct from the Saxon stump. When the corn is cut these bivouac fires go out, and the camp disappears, but the white ashes remain, and next season the smoke will rise again.

The barn here with its broad red roof, and the rickyard with the stone staddles, and the litter of chaff and straw, is the central rendezvous all the year of the resident labourers. Day by day, and at all hours, there is sure to be some of them about the place. The stamp of the land is on them. They border on the city, but are as distinctly agricultural and as immediately recognisable as in the heart of the country. This sturdy carter, as he comes round the corner of the straw-rick, cannot be mistaken.

He is short and thickly set, a man of some fifty years, but hard and firm of make. His face is broad and red, his shiny fat cheeks almost as prominent as his stumpy nose, likewise red and shiny. A fringe of reddish whiskers surrounds his chin like a cropped hedge. The eyes are small and set deeply, a habit of half-closing the lids when walking in the teeth of the wind and rain has caused them to appear still smaller. The wrinkles at the corners and the bushy eyebrows are more visible and pronounced than the eyes themselves, which are mere bright grey points twinkling with complacent good humour.

These red cheeks want but the least motion to break into a smile; the action of opening the lips to speak is sufficient to give that expression. The fur cap he wears allows the round shape of his head to be seen, and the thick neck which is the colour of a brick. He trudges deliberately round the straw-rick: there is something in the style of the man which exactly corresponds to the barn, and the straw, and the stone staddles, and the waggons. Could we look back three hundred years, just such a man would be seen in the midst of the same surroundings, deliberately trudging round the straw-ricks of Elizabethan days, calm and complacent though the Armada be at hand. There are the ricks just the same, here is the barn, and the horses are in good case; the wheat is coming on well. Armies may march, but these are the same.

When his waggon creaks along the road towards the town his eldest lad walks proudly by the leader's head, and two younger boys ride in the vehicle. They pass under the great elms; now the sunshine and now the shadow falls upon them; the horses move with measured step and without haste, and both horses and human folks are content in themselves.

As you sit in summer on the beach and gaze afar over the blue waters scarcely flecked with foam, how slowly the distant ship moves along the horizon. It is almost, but not quite, still. You go to lunch and return, and the vessel is still there; what patience the man at the wheel must have. So, now, resting here on the stile, see the plough yonder, travelling as it were with all sails set.

Three shapely horses in line draw the share. The traces are taut, the swing-tree like a yard braced square, the helmsman at the tiller bears hard upon the stilts. But does it move? The leading horse, seen distinct against the sky, lifts a hoof and places it down again, stepping in the last furrow made. But then there is a perceptible pause before the next hoof rises, and yet again a perceptible delay in the pull of the muscles. The stooping ploughman walking in the new furrow, with one foot often on the level and the other in the hollow, sways a little with the lurch of his implement, but barely drifts ahead.

While watched they scarcely move; but now look away for a time and on returning the plough itself and the lower limbs of the ploughman and the horses are out of sight. They have gone over a slope, and are "hull down"; a few minutes more, and they disappear behind the ridge. Look away again and read or dream, as you would on the beach, and then, see, the head and shoulders of the leading horse are up, and by-and-by the plough rises, as they come back on the opposite tack. Thus the long hours slowly pass.

Intent day after day upon the earth beneath his feet or upon the tree in the hedge yonder, by which, as by a lighthouse, he strikes out a straight furrow, his mind absorbs the spirit of the land. When the plough pauses, as he takes out his bread and cheese in the corner of the field for luncheon, he looks over the low cropped hedge and sees far off the glitter of the sunshine on the glass roof of the Crystal Palace. The light plays and dances on it, flickering as on rippling water. But, though hard by, he is not of London. The horses go on again, and his gaze is bent down upon the furrow.

A mile or so up the road there is a place where it widens, and broad strips of sward run parallel on both sides. Beside the path, but just off it, so as to be no obstruction, an aged man stands watching his sheep. He has stood there so long that at last the restless sheep dog has settled down on the grass. He wears a white smock-frock, and leans heavily on his long staff, which he holds with both hands, propping his chest upon it. His face is set in a frame of white—white hair, white whiskers, short white beard. It is much wrinkled with years; but still has a hale and hearty hue.

The sheep are only on their way from one part of the farm to another, perhaps half a mile; but they have already been an hour, and will probably occupy another in getting there. Some are feeding steadily; some are in a gateway, doing nothing, like their pastor; if they were on the loneliest slope of the Downs he and they could not be more unconcerned. Carriages go past, and neither the sheep nor the shepherd turn to look.

Suddenly there comes a hollow booming sound—a roar, mellowed and subdued by distance, with a peculiar beat upon the ear, as if a wave struck the nerve and rebounded and struck again in an infinitesimal fraction of time—such a sound as can only bellow from the mouth of

75

cannon. Another and another. The big guns at Woolwich are at work. The shepherd takes no heed—neither he nor his sheep.

His ears must acknowledge the sound, but his mind pays no attention. He knows of nothing but his sheep. You may brush by him along the footpath and it is doubtful if he sees you. But stay and speak about the sheep, and instantly he looks you in the face and answers with interest.

Round the corner of the straw-rick by the red-roofed barn there comes another man, this time with smoke-blackened face, and bringing with him an odour of cotton waste and oil. He is the driver of a steam ploughing engine, whose broad wheels in summer leave their impression in the deep white dust of the roads, and in moist weather sink into the soil at the gateways and leave their mark as perfect as in wax. But though familiar with valves, and tubes, and gauges, spending his hours polishing brass and steel, and sometimes busy with spanner and hammer, his talk, too, is of the fields.

He looks at the clouds, and hopes it will continue fine enough to work. Like many others of the men who are employed on the farms about town he came originally from a little village a hundred miles away, in the heart of the country. The stamp of the land is on him, too.

Besides the Irish, who pass in gangs and generally have a settled destination, many agricultural folk drift along the roads and lanes searching for work. They are sometimes alone, or in couples, or they are a man and his wife, and carry hoes. You can tell them as far as you can see them, for they stop and look over every gateway to note how the crop is progressing, and whether any labour is required.

On Saturday afternoons, among the crowd of customers at the shops in the towns, under the very shadow of the almost palatial villas of wealthy "City" men, there may be seen women whose dress and talk at once mark them out as agricultural. They have come in on foot from distant farms for a supply of goods, and will return heavily laden. No town-bred woman, however poor, would dress so plainly as these cottage matrons. Their daughters who go with them have caught the finery of the town, and they do not mean to stay in the cottage.

There is a bleak arable field, on somewhat elevated ground, not very far from the same old barn. In the corner of this field for the last two or three years a great pit of roots has been made: that is, the roots are piled together and covered with straw and earth. When this mound is opened in the early spring a stout, elderly woman takes her seat beside it, billhook in hand, and there she sits the day through trimming the roots one by one, and casting those that she has prepared aside ready to be carted away to the cattle.

A hurdle or two propped up with stakes, and against which some of the straw from a mound has been thrown, keeps off some of the wind. But the easterly breezes sweeping over the bare upland must rush round and over that slight bulwark with force but little broken. Holding the root in the left hand, she turns it round and slashes off the projections with quick blows, which seem to only just miss her fingers, laughing and talking the while with two children who have brought her some refreshment, and who roll and tumble and play about her. The scene might be bodily removed and set down a hundred miles away, in the midst of a western county, and would there be perfectly at one with the surroundings.

Here, as she sits and chops, the east wind brings the boom of trains continually rolling over an iron bridge to and from the metropolis. She was there two successive seasons to my knowledge; she, too, had the stamp of the land upon her.

The broad sward where the white-haired shepherd so often stands watching his sheep feeding along to this field, is decked in summer with many flowers. By the hedge the agrimony frequently lifts its long stem, surrounded with small yellow petals. One day towards autumn I noticed a man looking along a hedge, and found that he was gathering this plant. He had a small armful of the straggling stalks, from which the flowers were then fading. The herb had once a medicinal reputation, and, curious to know if it was still remembered, I asked him the name of the herb and what it was for. He replied that it was agrimony. "We makes tea of it, and it is good for the flesh," or, as he pronounced it, "fleysh."

WHEATFIELDS

The cornfields immediately without London on the southern side are among the first to be reaped. Regular as if clipped to a certain height, the level wheat shows the slope of the ground, corresponding to it, so that the glance travels swiftly and unchecked across the fields. They scarce seemed divided, for the yellow ears on either side rise as high as the cropped hedge between.

Red spots, like larger poppies, now appear above and now dive down again beneath the golden surface. These are the red caps worn by some of the reapers; some of the girls, too, have a red scarf across the shoulder or round the waist. By instinctive sympathy the heat of summer requires the contrast of brilliant hues, of scarlet and gold, of poppy and wheat.

A girl, as she rises from her stooping position, turns a face, brown, as if stained with walnut juice, towards me, the plain gold ring in her brown ear gleams, so, too, the rings on her finger, nearly black from the sun, but her dark eyes scarcely pause a second on a stranger. She is too busy, her tanned fingers are at work again gathering up the cut wheat. This is no gentle labour, but "hard hand-play," like that in the battle of the olden time sung by the Saxon poet.

The ceaseless stroke of the reaping-hook falls on the ranks of the corn: the corn yields, but only inch by inch. If the burning sun, or thirst, or weariness forces the reaper to rest, the fight too stays, the ranks do not retreat, and victory is only won by countless blows. The boom of a bridge as a train rolls over the iron girders resounds, and the brazen dome on the locomotive is visible for a moment as it passes across the valley. But no one heeds it—the train goes on its way to the great city, the reapers abide by their labour. Men and women, lads and girls, some mere children, judged by their stature, are plunged as it were in the wheat.

The few that wear bright colours are seen: the many who do not are unnoticed. Perhaps the dusky girl here with the red scarf may have some strain of the gipsy, some far-off reminiscence of the sunlit East which caused her to wind it about her. The sheaf grows under her fingers, it is bound about with a girdle of twisted stalks, in which mingle the green bine of convolvulus and the pink-streaked bells that must fade.

Heat comes down from above; heat comes up from beneath, from the dry, white earth, from the rows of stubble, as if emitted by the endless tubes of cut stalks pointing upwards. Wheat is a plant of the sun: it loves the heat, and heat crackles in the rustle of the straw. The pimpernels above which the hook passed are wide open: the larger white convolvulus trumpets droop languidly on the low hedge: the distant hills are dim with the vapour of heat; the very clouds which stay motionless in the sky reflect a yet more brilliant light from their white edges. Is there no shadow?

There is no tree in the field, and the low hedge can shelter nothing; but bordering the next, on rather higher ground, is an ash copse, with some few spruce firs. Resting on a rail in the shadow of these firs, a light air now and again draws along beside the nut-tree bushes of the hedge, the cooler atmosphere of the shadow, perhaps causes it. Faint as it is, it sways the heavy laden brome grass, but is not strong enough to lift a ball of thistledown from the bennets among which it is entangled.

How swiftly the much-desired summer comes upon us! Even with the reapers at work before one it is difficult to realise that it has not only come, but will soon be passing away. Sweet summer is but just long enough for the happy loves of the larks. It seems but yesterday, it is really more than five months since, that, leaning against the gate there, I watched a lark and his affianced on the ground among the grey stubble of last year still standing.

His crest was high and his form upright, he ran a little way and then sang, went on again and sang again to his love, moving parallel with him. Then passing from the old dead stubble to fresh-turned furrows, still they went side by side, now down in the valley between the clods, now mounting the ridges, but always together, always with song and joy, till I lost them across the brown earth. But even then from time to time came the sweet voice, full of hope in coming summer.

The day declined, and from the clear, cold sky of March the moon looked down, gleaming on the smooth planed furrow which the plough had passed. Scarce had she faded in the dawn ere the lark sang again, high in the morning sky. The evenings became dark; still he rose above the shadows and the dusky earth, and his song fell from the bosom of the night. With full untiring choir the joyous host heralded the birth of the corn; the slender forceless seed-leaves which came gently up till they had risen above the proud crests of the lovers.

Time advanced and the bare mounds about the field, carefully cleaned by the husbandman, were covered again with wild herbs and plants, like a fringe to a garment of pure green. Parsley and "gix," and clogweed, and sauce-alone, whose white flowers smell of garlic if crushed in the fingers, came up along the hedge; by the gateway from the bare trodden earth appeared the shepherd's purse; small must be the coin to go in its seed capsule, and therefore it was so called with grim and truthful humour, for the shepherd, hard as is his work, facing wind and weather, carries home but little money.

Yellow charlock shot up faster and shone bright above the corn; the oaks showered down their green flowers like moss upon the ground; the tree-pipits sang on the branches and descending to the wheat. The rusty chain-harrow, lying inside the gate, all tangled together, was concealed with grasses. Yonder the magpies fluttered over the beans among which they are always searching in spring; blackbirds, too, are fond of a beanfield.

Time advanced again, and afar on the slope bright yellow mustard flowered, a hill of yellow behind the elms. The luxuriant purple of trifolium, acres of rich colour, glowed in the sunlight. There was a scent of flowering beans, the vetches were in flower, and the peas which clung together for support—the stalk of the pea goes through the leaf as a painter thrusts his thumb through his palette. Under the edge of the footpath through the wheat a wild pansy blooms.

Standing in the gateway beneath the shelter of the elms as the clouds come over, it is pleasant to hear the cool refreshing rain come softly down; the green wheat drinks it as it falls, so that hardly a drop reaches the ground, and to-morrow it will be as dry as ever. Wood-pigeons call from the hedges, and blackbirds whistle in the trees; the sweet delicious rain refreshes them as it does the corn.

Thunder mutters in the distance, and the electric atmosphere rapidly draws the wheat up higher. A few days' sunshine and the first wheatear appears. Very likely there are others near, but standing with their hood of green leaf towards you, and therefore hidden. As the wheat comes into ear it is garlanded about with hedges in full flower.

It is midsummer, and midsummer, like a bride, is decked in white. On the high-reaching briars white June roses; white flowers on the lowly brambles; broad white umbels of elder in the corner, and white cornels blooming under the elm; honeysuckle hanging creamy white coronals round the ash boughs; white meadow-sweet flowering on the shore of the ditch; white clover, too, beside the gateway. As spring is azure and purple, so midsummer is white, and autumn golden. Thus the coming out of the wheat into ear is marked and welcomed with the purest colour.

But these, though the most prominent along the hedge, are not the only flowers; the prevalent white is embroidered with other hues. The brown feathers of a few reeds growing where the furrows empty the showers into the ditch, wave above the corn. Among the leaves of mallow its mauve petals are sheltered from the sun. On slender stalks the yellow vetchling blooms, reaching ambitiously as tall as the lowest of the brambles. Bird's-foot lotus, with red claws, is overtopped by the grasses.

The elm has a fresh green—it has put forth its second or midsummer shoot; the young leaves of the aspen are white, and the tree as the wind touches it seems to turn grey. The furrows run to the ditch under the reeds, the ditch declines to a little streamlet which winds all hidden by willowherb and rush and flag, a mere trickle of water under brooklime, away at the feet of the corn. In the shadow, deep down beneath the crumbling bank which is only held up by the roots of the grasses, is a forget-me-not with a tiny circlet of yellow in the centre of its petals.

The coming of the ears of wheat forms an era and a date, a fixed point in the story of the summer. It is then that, soon after dawn, the clear sky assumes the delicate and yet luscious purple which seems to shine through the usual atmosphere, as if its former blue became translucent and an inner and ethereal light of colour was shown. As the sun rises higher the brilliance of his rays overpowers it, and even at midsummer it is but rarely seen.

The morning sky is often, too, charged with saffron, or the blue is clear, but pale, and the sunrise might be watched for many mornings without the appearance of this exquisite hue. Once seen, it will ever be remembered. Upon the Downs in early autumn, as the vapours clear away, the same colour occasionally gleams from the narrow openings of blue sky. But at midsummer, above the opening wheatears, the heaven from the east to the zenith is flushed with it.

At noonday, as the light breeze comes over, the wheat rustles the more because the stalks are stiffening and swing from side to side from the root instead of yielding up the stem. Stay now at every gateway and lean over while the midsummer hum sounds above. It is a peculiar sound, not like the querulous buzz of the honey, nor the drone of the humble bee, but a sharp ringing resonance like that of a tuning-fork. Sometimes, in the far-away country where it is often much louder, the folk think it has a threatening note.

Here the barley has taken a different tint now the beard is out; here the oats are straggling forth from their sheath; here a pungent odour of mustard in flower comes on the air; there a poppy faints with broad petals flung back and drooping, unable to uphold its gorgeous robes. The flower of the field pea, here again, would make a model for a lady's hat; so would a butterfly with closed wings on the verge of a leaf; so would the broom blossom, or the pink flower of the restharrow. This hairy caterpillar, creeping along the hawthorn, which if touched, immediately coils itself in a ring, very recently was thought a charm in distant country places for some diseases of childhood, if hung about the neck. Hedge mustard, yellow and ragged and dusty, stands by the gateway.

In the evening, as the dew gathers on the grass, which feels cooler to the hand some time before an actual deposit, the clover and vetches close their leaves—the signal the hares have been waiting for to venture from the sides of the fields where they have been cautiously roaming, and take bolder strolls across the open and along the lanes. The aspens rustle louder in the stillness of the evening; their leaves not only sway to and fro, but semi-rotate upon the stalks, which causes their scintillating appearance. The stars presently shine from the pale blue sky, and the wheat shimmers dimly white beneath them.

So time advances till to-day, watching the reapers from the shadow of the copse, it seems as if within that golden expanse there must be something hidden, could you but rush in quickly and seize it—some treasure of the sunshine; and there *is* a treasure, the treasure of life stored in those little grains, the slow product of the sun. But it cannot be grasped in an impatient moment—it must be gathered with labour. I have threshed out in my hand three ears of the ripe wheat: how many foot-pounds of human energy do these few light grains represent?

The roof of the Crystal Palace yonder gleams and sparkles this afternoon as if it really were crystal under the bright rays. But it was concealed by mist when the ploughs in the months gone by were guided in these furrows by men, hard of feature and of hand, stooping to their toil. The piercing east wind scattered the dust in clouds, looking at a distance like small rain across the field, when grey-coated men, grey too of beard, followed the red drill to and fro.

How many times the horses stayed in this sheltered corner while the ploughmen and their lads ate their crusts! How many times the farmer and the bailiff, with hands behind their backs, considering, walked along the hedge taking counsel of the earth if they had done right! How many times hard gold and silver was paid over at the farmer's door for labour while yet the plant was green; how many considering cups of ale were emptied in planning out the future harvest!

Now it is come, and still more labour—look at the reapers yonder— and after that more time and more labour before the sacks go to the market. Hard toil and hard fare: the bread which the reapers have brought with them for their luncheon is hard and dry, the heat has dried it like a chip. In the corner of the field the women have gathered some sticks and lit a fire—the flame is scarce seen in the sunlight, and the sticks seem eaten away as they burn by some invisible power. They are boiling a kettle, and their bread, too, which they will soak in the tea, is dry and chip-like. Aside, on the ground by the hedge, is a handkerchief tied at the corners, with a few mushrooms in it.

The scented clover field—the white campions dot it here and there— yields a rich, nectareous food for ten thousand bees, whose hum comes together with its odour on the air. But these men and women and children ceaselessly toiling know no such sweets; their food is as hard as their labour. How many foot-pounds, then, of human energy do these grains in my hand represent? Do they not in their little compass contain the potentialities, the past and the future, of human life itself?

Another train booms across the iron bridge in the hollow. In a few hours now the carriages will be crowded with men hastening home from their toil in the City. The narrow streak of sunshine which day by day falls for a little while upon the office floor, yellowed by the dingy pane, is all, perhaps, to remind them of the sun and sky, of the forces of nature; and that little is unnoticed. The pressure of business is so severe in these later days that in the hurry and excitement it is not wonderful many should forget that the world is not comprised in the court of a City thoroughfare.

Rapt and absorbed in discount and dollars, in bills and merchandise, the over-strung mind deems itself all—the body is forgotten, the physical body, which is subject to growth and change, just as the plants and the very grass of the field. But there is a subtle connection between the physical man and the great nature which comes pressing up so closely to the metropolis. He still depends in the nineteenth century, as in the dim ages before the Pyramids, upon this tiny yellow grain here, rubbed out from the ear of wheat. The clever mechanism of the locomotive which bears him to and fro, week after week and month after month, from home to office and from office home, has not rendered him in the least degree independent of this.

But it is no wonder that these things are forgotten in the daily struggle of London. And if the merchant spares an abstracted glance from the morning or evening newspaper out upon the fields from the carriage window, the furrows of the field can have but little meaning. Each looks to him exactly alike. To the farmers and the labourer such and such a furrow marks an acre and has its bearing, but to the passing glance it is not so. The work in the field is so slow; the passenger by rail sees, as it seems to him, nothing going on; the corn may sow itself almost for all that is noteworthy in apparent labour.

Thus it happens that, although the cornfields and the meadows come so closely up to the offices and warehouses of mighty London, there is a line and mark in the minds of men between them; the man of merchandise does not see what the man of the fields sees, though both may pass the same acres every morning. It is inevitable that it should be so. It is easy in London to forget that it is midsummer, till, going some day into Covent Garden Market, you see baskets of the cornflower, or blue-bottle as it is called in the country, ticketed "Corinne," and offered for sale. The lovely azure of the flower recalls the scene where it was first gathered long since at the edge of the wheat.

By the copse here now the teazles lift their spiny heads high in the hedge, the young nuts are browning, the wild mints flowering on the shores of the ditch, and the reapers are cutting ceaselessly at the ripe corn. The larks have brought their loves to a happy conclusion. Besides them the wheat in its day has sheltered many other creatures—both animals and birds.

Hares raced about it in the spring, and even in the May sunshine might be seen rambling over the slopes. As it grew higher it hid the leverets and the partridge chicks. Toll has been taken by rook, and sparrow, and pigeon. Enemies, too, have assailed it; the daring couch invaded it, the bindweed climbed up the stalk, the storm rushed along and beat it down. Yet it triumphed, and to-day the full sheaves lean against each other.

THE CROWS

On one side of the road immediately after quitting the suburb there is a small cover of furze. The spines are now somewhat browned by the summer heats, and the fern which grows about every bush trembles on the balance of colour between green and yellow. Soon, too, the tall wiry grass will take a warm brown tint, which gradually pales as the autumn passes into winter, and finally bleaches to greyish white.

Looking into the furze from the footpath, there are purple traces here and there at the edge of the fern where the heath-bells hang. On a furze branch, which projects above the rest, a furze chat perches, with yellow blossom above and beneath him. Rushes mark the margin of small pools and marshy spots, so overhung with brambles and birch branches, and so closely surrounded by gorse, that they would not otherwise be noticed.

But the thick growth of rushes intimates that water is near, and upon parting the bushes a little may be seen, all that has escaped evaporation in the shade. From one of these marshy spots I once—and once only—observed a snipe rise, and after wheeling round return and settle by another. As the wiry grass becomes paler with the fall of the year, the rushes, on the contrary, from green become faintly yellow, and presently brownish. Grey grass and brown rushes, dark furze, and fern, almost copper in hue from frost, when lit up by a gleam of winter sunshine form a pleasant breadth of warm colour in the midst of bare fields.

After continuous showers in spring, lizards are often found in the adjacent gardens, their dark backs as they crawl over the patches being almost exactly the tint of the moist earth. If touched, the tail is immediately coiled, the body stiffens, and the creature appears dead. They are popularly supposed to come from the furze, which is also believed to shelter adders.

There is, indeed, scarcely a cover in Surrey and Kent which is not said to have its adders; the gardeners employed at villas close to the metropolis occasionally raise an alarm, and profess to have seen a viper in the shrubberies, or the ivy, or under an old piece of bast. Since so few can distinguish at a glance between the common snake and the adder it is as well not to press too closely upon any reptile that may chance to be heard rustling in the grass, and to strike tussocks with the walking-stick before sitting down to rest, for the adder is only dangerous when unexpectedly encountered.

In the roadside ditch by the furze the figwort grows, easily known by its coarse square stem; and the woody bines, if so they may be called, or stalks of bitter-sweet, remain all the winter standing in the hawthorn hedge. The first frosts, on the other hand, shrivel the bines of white bryony, which part and hang separated, and in the spring a fresh bine pushes up with greyish green leaves and tendrils feeling for support. It is often observed that the tendrils of this bryony coil both ways, with and against the sun.

But it must be remembered in looking for this that it is the same tendril which should be examined, and not two different ones. It will then be seen that the tendril, after forming a spiral one way, lengthens out like a tiny green wax taper, and afterwards turns the other. Sometimes it resumes the original turn before reaching a branch to cling to, and may thus be said to have revolved in three directions. The dusty celandine grows under the bushes; and its light green leaves seem to retain the white dust from the road. Ground ivy creeps everywhere over the banks, and covers the barest spot. In April its flowers, though much concealed by leaves, dot the sides of the ditches with colour, like the purple tint that lurks in the amethyst.

A small black patch marks the site of one of those gorse fires which are so common in Surrey. This was extinguished before it could spread beyond a few bushes. The crooked stems remain black as charcoal, too much burnt to recover, and in the centre a young birch scorched by the flames stands leafless. This barren birch, bare of foliage and apparently unattractive, is the favourite resort of yellow-hammers. Perching on a branch towards evening a yellow-hammer will often sit and sing by the hour together, as if preferring to be clear of leafy sprays.

The somewhat dingy hue of many trees as the summer begins to wane is caused not only by the fading of the green, but by the appearance of spots upon the leaves, as may be seen on those birches which grow among the furze. But in spring and early summer their fresh light green contrasts with masses of bright yellow gorse bloom. Just before then—just as the first leaves are opening—the chiffchaffs come.

The first spring I had any knowledge of this spot was mild, and had been preceded by mild seasons. The chiffchaffs arrived all at once, as it seemed, in a bevy, and took possession of every birch about the furze, calling incessantly with might and main. The willow-wrens were nearly as numerous. All the gorse seemed full of them for a few days. Then by degrees they gradually spread abroad, and dispersed among the hedges.

But in the following springs nothing of the kind occurred. Chiffchaff and willow-wren came as usual, but they did not arrive in a crowd at once. This may have been owing to the flight going elsewhere, or possibly the flock were diminished by failure to rear the young broods in so drenching a season as 1879, which would explain the difference observed next spring. There was no scarcity, but there was a lack of the bustle and excitement and flood of song that accompanied their advent two years before.

Upon a piece of waste land at the corner of the furze a very large cinder and dust heap was made by carting refuse there from the neighbouring suburb. During the sharp and continued frosts of the winter this dust-heap was the resort of almost every species of bird—sparrows, starlings, greenfinches, and rooks searching for any stray morsels of food. Some birdcatchers soon noticed this concourse, and spread their nets among the adjacent rushes, but fortunately with little success.

I say fortunately, not because I fear the extinction of small birds, but because of the miserable fate that awaits the captive. Far better for the frightened little creature to have its neck at once twisted and to die than to languish in cages hardly large enough for it to turn in behind the dirty panes of the windows in the Seven Dials.

The happy greenfinch—I use the term of forethought, for the greenfinch seems one of the very happiest of birds in the hedges— accustomed during all its brief existence to wander in company with friends from bush to bush, and tree to tree, must literally pine its heart out. Or it may be streaked with bright paint and passed on some unwary person for a Java sparrow or a "blood-heart."

The little boy who dares to take a bird's nest is occasionally fined and severely reproved. The ruffian-like crew who go forth into the pastures and lanes about London, snaring and netting full-grown birds by the score, are permitted to ply their trade unchecked. I mean to say that there is no comparison between the two things. An egg has not yet advanced to consciousness or feeling: the old birds, if their nest is taken, frequently build another. The lad has to hunt for the nest, to climb for it or push through thorns, and may be pricked by brambles and stung by nettles. In a degree there is something to him approaching to sport in nesting.

But these birdcatchers simply stand by the ditch with their hands in their pockets sucking a stale pipe. They would rather lounge there in the bitterest north-east wind that ever blew than do a single hour's honest work. Blackguard is written in their faces. The poacher needs some courage, at least; he knows a penalty awaits detection. These fellows have no idea of sport, no courage, and no skill, for their tricks are simplicity itself, nor have they the pretence of utility, for they do not catch birds for the good of the farmers or the market gardeners, but merely that they may booze without working for the means.

Pity it is that any one can be found to purchase the product of their brutality. No one would do so could they but realise the difference to the captive upon which they are lavishing their mistaken love, between the cage, the alternately hot and cold room (as the fire goes out at night), the close atmosphere and fumes that lurk near the ceiling, and the open air and freedom to which it was born.

The rooks only came to the dust-heap in hard weather, and ceased to visit it so soon as the ground relaxed and the ploughs began to move. But a couple of crows looked over the refuse once during the day for months till men came to sift the cinders. These crows are permanent residents. Their rendezvous is a copse, only separate from the furze by the highway.

They are always somewhere near, now in the ploughed fields, now in the furze, and during the severe frost of last winter in the road itself, so sharply driven by hunger as to rise very unwillingly on the approach of passengers. A meadow opposite the copse is one of their favourite resorts. There are anthills, rushes, and other indications of not too rich a soil in this meadow, and in places the prickly restharrow grows among the grass, bearing its pink flower in summer. Perhaps the coarse grass and poor soil are productive of grubs and insects, for not only the crows, but the rooks, continually visit it.

One spring, hearing a loud chattering in the copse, and recognising the alarm notes of the missel-thrush, I cautiously crept up the hedge, and presently found three crows up in a birch tree, just above where the thrushes were calling. The third crow—probably a descendant of the other two—had joined in a raid upon the missel-thrushes' brood. Both defenders and assailants were in a high state of excitement; the thrushes screeching, and the crows in a row one above the other on a branch, moving up and down it in a restless manner. I fear they had succeeded in their purpose, for no trace of the young birds was visible.

The nest of the missel-thrush is so frequently singled out for attack by crows that it would seem the young birds must possess a peculiar and attractive flavour; or is it because they are large? There are more crows round London than in a whole county, where the absence of manufactures and the rural quiet would seem favourable to bird life. The reason, of course, is that in the country the crows frequenting woods are shot and kept down as much as possible by gamekeepers.

In the immediate environs of London keepers are not about, and even a little farther away the land is held by many small owners, and game preservation is not thought of. The numerous pieces of waste ground, "to let on building lease," the excavated ground, where rubbish can be thrown, the refuse and ash heaps—these are the haunts of the London crow. Suburban railway stations are often haunted by crows, which perch on the telegraph wires close to the back windows of the houses that abut upon the metals. There they sit, grave and undisturbed by the noisy engines which pass beneath them.

In the shrubberies around villa gardens, or in the hedges of the small paddocks attached, thrushes and other birds sometimes build their nests. The children of the household watch the progress of the nest, and note the appearance of the eggs with delight. Their friends of larger growth visit the spot occasionally, and orders are given that the birds shall be protected, the gardeners become gamekeepers, and the lawn or shrubbery is guarded like a preserve. Everything goes well till the young birds are almost ready to quit the nest, when one morning they are missing.

The theft is, perhaps, attributed to the boys of the neighbourhood, but unjustly, unless plain traces of entry are visible. It is either cats or crows. The cats cannot be kept out, not even by a dog, for they watch till his attention is otherwise engaged. Food is not so much the object as the pleasure of destruction, for cats will kill and yet not eat their victim. The crow may not have been seen in the garden, and it may be said that he could not have known of the nest without looking round the place. But the crow is a keen observer, and has not the least necessity to search for the nest.

He merely keeps a watch on the motions of the old birds of the place, and knows at once by their flight being so continually directed to one spot that there there their treasure lies. He and his companion may come very early in the morning—summer mornings are bright as noonday long before the earliest gardener is abroad—or they may come in the dusk of the evening. Crows are not so particular in retiring regularly to roost as the rook.

The furze and copse frequented by the pair which I found attacking the missel-thrushes are situate at the edge of extensive arable fields. In these, though not overlooked by the gamekeepers, there is a good deal of game which is preserved by the tenants of the farm. After the bitter winter and wet summer of 1879, there was a complaint, too well founded, that the partridges were diminished in numbers. But the crows were not. There were as many of them as ever. When there were many partridges the loss of a few eggs or chicks was not so important. But when there are but few, every egg or chick destroyed retards the re-stocking of the fields.

The existence of so many crows all round London is, in short, a constant check upon the game. The belt of land immediately outside the houses, and lying between them and the plantations which are preserved, is the crow's reserve, where he hunts in security. He is so safe that he has almost lost all dread of man, and his motions can be observed without trouble. The ash-heap at the corner of the furze, besides the crows, became the resort of rats, whose holes were so thick in the bank as to form quite a bury. After the rats came the weasels.

When the rats were most numerous, before the ash-heap was sifted, there was a weasel there nearly every day, slipping in and out of their holes. In the depth of the country an observer might walk some considerable distance and wait about for hours without seeing a weasel; but here by the side of a busy suburban road there were plenty. Professional ratcatchers ferreted the bank once or twice, and filled their iron cages. With these the dogs kept by dog-fanciers in the adjacent suburb were practised in destroying vermin at so much a rat. Though ferreted and hunted down by the weasels the rats were not rooted out, but remained till the ash-heap was sifted and no fresh refuse deposited.

In one place among the gorse, the willows, birches, and thorn bushes make a thick covert, which is adjacent to several of the hidden pools previously mentioned. Here a brook-sparrow or sedge-reedling takes up his quarters in the spring, and chatters on, day and night, through the summer. Visitors to the opera and playgoers returning in the first hours of the morning from Covent Garden or Drury Lane can scarcely fail to hear him if they pause but one moment to listen to the nightingale.

The latter sings in one bush and the sedge-reedling in another close together. The moment the nightingale ceases the sedge-reedling lifts his voice, which is a very penetrating one, and in the silence of the night may be heard some distance. This bird is credited with imitating the notes of several others, and has been called the English mocking-bird, but I strongly doubt the imitation. Nor, indeed, could I ever trace the supposed resemblance of its song to that of other birds.

It is a song of a particularly monotonous character. It is distinguishable immediately, and if the bird happens to nest near a house, is often disliked on account of the loud iteration. Perhaps those who first gave it the name of the mocking-bird were not well acquainted with the notes of the birds which they fancied it to mock. To mistake it for the nightingale, some of whose tones it is said to imitate, would be like confounding the clash of cymbals with the soft sound of a flute.

Linnets come to the furze, and occasionally magpies, but these latter only in winter. Then, too, golden-crested wrens may be seen searching in the furze bushes, and creeping round and about the thorns and brambles. There is a roadside pond close to the furze, the delight of horses and cattle driven along the dusty way in summer. Along the shelving sandy shore the wagtails run, both the pied and the yellow, but few birds come here to wash; for that purpose they prefer a running stream if it be accessible.

Upon the willow trees which border it, a reed-sparrow or blackheaded bunting may often be observed. One bright March morning, as I came up the road, just as the surface of the pond became visible it presented a scene of dazzling beauty. At that distance only the tops of the ripples were seen, reflecting the light at a very low angle. The result was that the eye saw nothing of the water or the wavelet, but caught only the brilliant glow. Instead of a succession of sparkles there seemed to be a golden liquid floating on the surface as oil floats—a golden liquid two or three inches thick, which flowed before the wind.

Besides this surface of molten gold there was a sheen and flicker above it, as if a spray or vapour, carried along, or the crests of the wavelets blown over, was also of gold. But the metal conveys no idea of the glowing, lustrous light which filled the hollow by the dusty road. It was visible from one spot only, a few steps altering the angle lessened the glory, and as the pond itself came into view there was nothing but a ripple on water somewhat thick with suspended sand. Thus things change their appearance as they are looked at in different ways.

A patch of water crowsfoot grows on the farthest side of the pond, and in early summer sends up lovely white flowers.

HEATHLANDS

Sandown has become one of the most familiar places near the metropolis, but the fir woods at the back of it are perhaps scarcely known to exist by many who visit the fashionable knoll. Though near at hand, they are shut off by the village of Esher; but a mile or two westwards, down the Portsmouth highway, there is a cart road on the left hand which enters at once into the woods.

The fine white sand of the soil is only covered by a thin coating of earth formed from the falling leaves and decayed branches, so thin that it may sometimes be rubbed away by the foot or even the fingers. Grass and moss grow sparingly in the track, but wherever wheels or footsteps have passed at all frequently the sand is exposed in white streaks under the shadowy firs. In grass small objects often escape observation, but on such a bare surface everything becomes visible. Coming to one of these places on a summer day, I saw a stream of insects crossing and recrossing, from the fern upon one side to the fern upon the other.

They were ants, but of a very much larger species than the little red-and-black "emmets" which exist in the meadows. These horse ants were not much less than half an inch in length, with a round spot at each end like beads, or the black top of long pins. The length of their legs enabled them to move much quicker, and they raced to and fro over the path with great rapidity. The space covered by the stream was a foot or more broad, all of which was crowded and darkened by them, and as there was no cessation in the flow of this multitude, their numbers must have been immense.

Standing a short way back, so as not to interfere with their proceedings, I saw two of these insects seize hold of a twig, one at each end. The twig, which was dead and dry, and had dropped from a fir, was not quite so long as a match, but rather thicker. They lifted this stick with ease, and carried it along, exactly as labourers carry a plank. A few short blades of grass being in the way they ran up against them, but stepped aside, and so got by. A cart which had passed a long while since had forced down the sand by the weight of its load, leaving a ridge about three inches high, the side being perpendicular.

Till they came to this cliff the two ants moved parallel, but here one of them went first, and climbed up the bank with its end of the stick, after which the second followed and brought up the other. An inch or two farther, on the level ground, the second ant left hold and went away, and the first laboured on with the twig and dragged it unaided across the rest of the path. Though many other ants stayed and looked at the twig a moment, none of them now offered assistance, as if the chief obstacle had been surmounted.

Several other ants passed, each carrying the slender needles which fall from firs, and which seemed nothing in their powerful grasp. These burdens of wood all went in one direction, to the right of the path.

I took a step there, but stayed to watch two more ants, who had got a long scarlet fly between them, one holding it by the head and the other by the tail. They were hurrying their prey over the dead leaves and decayed sticks which strewed the ground, and dragging it mercilessly through moss and grass. I put the tip of my stick on the victim, but instead of abandoning it they tugged and pulled desperately, as if they would have torn it to pieces rather than have yielded. So soon as I released it away they went through the fragments of branches, rushing the quicker for the delay.

A little farther there was a spot where the ground for a yard or two was covered with small dead brown leaves, last year's, apparently of birch, for some young birch saplings grew close by. One of these leaves

97

suddenly rose up and began to move of itself, as it seemed; an ant had seized it, and holding it by the edge travelled on, so that as the insect was partly hidden under it, the leaf appeared to move alone, now over sticks and now under them. It reminded me of the sight which seemed so wonderful to the early navigators when they came to a country where, as they first thought, the leaves were alive and walked about.

The ant with the leaf went towards a large heap of rubbish under the sapling birches. While watching the innumerable multitude of these insects, whose road here crossed these dead dry leaves, I became conscious of a rustling sound, which at first I attributed to the wind, but seeing that the fern was still, and that the green leaves of a Spanish chestnut opposite did not move, I began to realise that this creeping, rustling noise, distinctly audible, was not caused by any wind, but by the thousands upon thousands of insects passing over the dead leaves and among the grass. Stooping down to listen better, there could be no doubt of it: it was the tramp of this immense army.

The majority still moved in one direction, and I found it led to the heap of rubbish over which they swarmed. This heap was exactly what might have been swept together by half-a-dozen men using long gardeners' brooms, and industriously clearing the ground under the firs of the fragments which had fallen from them. It appeared to be entirely composed of small twigs, fir-needles, dead leaves, and similar things. The highest part rose about level with my chest—say, between four and five feet—the heap was irregularly circular, and not less than three or four yards across, with sides gradually sloping. In the midst stood the sapling birches, their stumps buried in it, the rubbish having been piled up around them.

This heap was, in fact, the enormous nest or hill of a colony of horse ants. The whole of it had been gathered together, leaf by leaf, and twig by twig, just as I had seen the two insects carrying the little stick, and the third the brown leaf above itself. It really seemed some way round the outer circumference of the nest, and while walking round it was necessary to keep brushing off the ants which dropped on the shoulder from the branches of the birches. For they were everywhere; every inch of ground, every bough was covered with them. Even standing near it was needful to kick the feet continually against the black stump of a fir which had been felled to jar them off, and this again brought still more, attracted by the vibration of the ground.

The highest part of the mound was in the shape of a dome, a dome whitened by layers of fir-needles, which was apparently the most recent part and the centre of this year's operations. The mass of the heap, though closely compacted, was fibrous, and a stick could be easily thrust into it, exposing the eggs. No sooner was such an opening made, and the stick withdrawn from the gap, than the ants swarmed into it, falling headlong over upon each other, and filling the bottom with their struggling bodies. Upon leaving the spot, to follow the footpath, I stamped my feet to shake down any stray insects, and then took off my coat and gave it a thorough shaking.

Immense ant-hills are often depicted in the illustrations to tropical travels, but this great pile, which certainly contained more than a cartload, was within a few miles of Hyde Park Corner. From nests like this large quantities of eggs are obtained for feeding the partridges hatched from the eggs collected by mowers and purchased by keepers. Part of the nest being laid bare with any tool, the eggs are hastily taken out in masses and thrown into a sack. Some think that ant's eggs, although so favourite a food, are not always the most advantageous. Birds which have been fed freely on these eggs become fastidious, and do not care for much else, so that if the supply fails they fall off in condition. If there are sufficient eggs to last the season, then a few every day produce the best effect; if not they had better not have a feast followed by a fast.

The sense of having a roof overhead is felt in walking through a forest of firs like this, because the branches are all at the top of the trunks. The stems rise to the same height, and then the dark foliage spreading forms a roof. As they are not very near together the eye can see some distance between them, and as there is hardly any underwood or bushes—nothing higher than the fern—there is a space open and unfilled between the ground and the roof so far above.

A vast hollow extends on every side, nor is it broken by the flitting of birds or the rush of animals among the fern. The sudden note of a wood-pigeon, hoarse and deep, calling from a fir-top, sounds still louder and ruder in the spacious echoing vault beneath, so loud as at first to resemble the baying of a hound. The call ceases, and another of these watch-dogs of the woods takes it up afar off.

There is an opening in the monotonous firs by some rising ground, and the sunshine falls on young Spanish chestnuts and underwood, through which is a little-used footpath. If firs are planted in wildernesses with the view of ultimately covering the barren soil with fertile earth, formed by the decay of vegetable matter, it is, perhaps, open to discussion as to whether the best tree has been chosen. Under firs the ground is generally dry, too dry for decay; the resinous emanations rather tend to preserve anything that falls there.

No underwood or plants and little grass grows under them; these, therefore, which make soil quickest, are prevented from improving the earth. The needles of firs lie for months without decay; they are, too, very slender, and there are few branches to fall. Beneath any other trees (such as the edible chestnut and birch, which seem to grow here), there are the autumn leaves to decay, the twigs and branches which fall off, while grasses and plants flourish, and brambles and underwood grow freely. The earth remains moist, and all these soon cause an increase of the fertility; so that, unless fir-tree timber is very valuable, and I never heard that it was, I would rather plant a waste with any other tree or brushwood, provided, of course, it would grow.

It is a pleasure to explore this little dell by the side of the rising ground, creeping under green boughs which brush the shoulders, after the empty space of the firs. Within there is a pond, where lank horsetails grow thickly, rising from the water. Returning to the rising ground I pursue the path, still under the shadow of the firs. There is no end to them—the vast monotony has no visible limit. The brake fern—it is early in July—has not yet reached its full height, but what that will be is shown by these thick stems which rise smooth and straight, fully three feet to the first frond.

A woodpecker calls, and the gleam of his green and gold is visible for a moment as he hastens away—the first bird, except the wood-pigeons, seen for an hour, yet there are miles of firs around. After a time the ground rises again, the tall firs cease, but are succeeded by younger firs. These are more pleasant because they do not exclude the sky. The sunshine lights the path, and the summer blue extends above. The fern, too, ceases, and the white sand is now concealed by heath, with here and there a dash of colour. Furze chats call, and flit to and fro; the hum of bees is heard once more—there was not one under the vacant shadow; and swallows pass overhead.

At last emerging from the firs the open slope is covered with heath only, but heath growing so thickly that even the narrow footpaths are hidden by the overhanging bushes of it. Some small bushes of furze here and there are dead and dry, but every prickly point appears perfect; when struck with the walking-stick the bush crumbles to pieces. Beneath and amid the heath what seems a species of lichen grows so profusely as to give a grey undertone. In places it supplants the heath, the ground is concealed by lichen only, which crunches under the foot like hoar-frost. Each piece is branched not unlike a stag's antlers; gather a handful and it crumbles to pieces in the fingers, dry and brittle.

A quarry for sand has been dug down some eight or ten feet, so that standing in it nothing else is visible. This steep scarp shows the strata, yellow sand streaked with thin brown layers; at the top it is fringed with heath in full flower, bunches of purple bloom overhanging the edge, and behind this the azure of the sky.

Here, where the ground slopes gradually, it is entirely covered with the purple bells; a sheen and gleam of purple light plays upon it. A fragrance of sweet honey floats up from the flowers where grey hive-bees are busy. Ascending still higher and crossing the summit, the ground almost suddenly falls away in a steep descent, and the entire hillside, seen at a glance, is covered with heath, and heath alone. A bunch at the very edge offers a purple cushion fit for a king; resting here a delicious summer breeze, passing over miles and miles of fields and woods yonder, comes straight from the distant hills. Along those hills the lines of darker green are woods; there are woods to the south, and west, and east, heath around and in the rear the gaze travels over the tops of the endless firs. But southwards is sweetest; below, beyond the verge of the heath, the corn begins, and waves in the wind. It is the breeze that makes the summer day so lovely.

The eggs of the nighthawk are sometimes found at this season near by. They are laid on the ground, on the barest spots, where there is no herbage. At dusk, the nighthawk wheels with a soft yet quick flight over the ferns and about the trees. Along the hedges bounding the heath butcher-birds watch for their prey—sometimes on the furze, sometimes on a branch of ash. Wood-sage grows plentifully on the banks by the roads; it is a plant somewhat resembling a lowly nettle; the leaves have a hop-like scent, and so bitter and strong is the odour that immediately after smelling them the mouth for a moment feels dry with a sense of thirst.

The angle of a field by the woods on the eastern side of the heath, the entire corner, is blue in July with viper's bugloss. The stalks rise some two feet, and are covered with minute brown dots; they are rough, and the lower part prickly. Blue flowers in pairs, with pink stamens and pink buds, bloom thickly round the top, and as each plant has several stalks, it is very conspicuous where the grass is short.

There are hundreds of these flowers in this corner, and along the edge of the wood; a quarter of an acre is blue with them. So indifferent are people to such things that men working in the same field, and who had pulled up the plant and described its root as like that of a dock, did not know its name. Yet they admired it. "It is an innocent-looking flower," they said, that is, pleasant to look at.

By the roadside I thought I saw something red under the long grass of the mound, and, parting the blades, found half-a-dozen wild strawberries. They were larger than usual, and just ripe. The wild strawberry is a little more acid than the cultivated, and has more flavour than would be supposed from its small size.

Descending to the lower ground again, the brake fills every space between the trees; it is so thick and tall that the cows which wander about, grazing at their will, each wear a bell slung round the neck, that their position may be discovered by sound. Otherwise it would be difficult to find them in the fern or among the firs. There are many swampy places here, which should be avoided by those who dislike snakes. The common harmless snakes are numerous in this part, and they always keep near water. They often glide into a mole's "angle," or hole, if found in the open.

Adders are known to exist in the woods round about, but are never, or very seldom, seen upon the heath itself. In the woods of the neighbourhood they are not uncommon, and are sometimes killed for the sake of the oil. The belief in the virtue of adder's fat, or oil, is still firm; among other uses it is considered the best thing for deafness, not, of course, resulting from organic defect. For deafness, the oil should be applied by pouring a small quantity into the ear, exactly in the same manner as in the play the poison is poured into the ear of the sleeping king. Cures are declared to be effected by this oil at the present day.

It is procured by skinning the adder, taking the fat, and boiling it; the result is a clear oil, which never thickens in the coldest weather. One of these reptiles on being killed and cut open was found to contain the body of a full-grown toad. The old belief that the young of the viper enters its mouth for refuge still lingers. The existence of adders in the woods here seems so undoubted that strangers should be a little careful if they leave the track. Viper's bugloss, which grows so freely by the heath, was so called because anciently it was thought to yield an antidote to the adder's venom.

THE RIVER

There is a slight but perceptible colour in the atmosphere of summer. It is not visible close at hand, nor always where the light falls strongest, and if looked at too long it sometimes fades away. But over gorse and heath, in the warm hollows of wheatfield, and round about the rising ground there is something more than air alone. It is not mist, nor the hazy vapour of autumn, nor the blue tints that come over the distant hills and woods.

As there is a bloom upon the peach and grape, so this is the bloom of summer. The air is ripe and rich, full of the emanations, the perfume, from corn and flower and leafy tree. In strictness the term will not, of course, be accurate, yet by what other word can this appearance in the atmosphere be described but as a bloom? Upon a still and sunlit summer afternoon it may be seen over the osier-covered islets in the Thames immediately above Teddington Lock.

It hovers over the level cornfields that stretch towards Richmond, and along the ridge of the wooded hills that bound them. The bank by the towing-path is steep and shadowless, being bare of trees or hedge; but the grass is pleasant to rest on, and heat is always more supportable near flowing water. In places the friable earth has crumbled away, and there, where the soil and the stones are exposed, the stonecrop flourishes. A narrow footpath on the summit, raised high above the water, skirts the corn, and is overhung with grass heavily laden by its own seed.

Sometimes in early June the bright trifolium, drooping with its weight of flower, brushes against the passer-by—acre after acre of purple. Occasionally the odour of beans in blossom floats out over the river.

Again, above the green wheat the larks rise, singing as they soar; or later on the butterflies wander over the yellow ears. Or, as the law of rotation dictates, the barley whitens under the sun. Still, whether in the dry day, or under the dewy moonlight, the plain stretching from the water to the hills is never without perfume, colour, or song.

There stood, one summer not long since, in the corner of a barley field close to the Lock, within a stone's throw, perfect shrubs of mallow, rising to the shoulder, thick as a walking-stick, and hung with flower. Poppies filled every interstice between the barley stalks, their scarlet petals turned back in very languor of exuberant colour, as the awns, drooping over, caressed them. Poppies, again, in the same fields formed a scarlet ground from which the golden wheat sprang up, and among it here and there shone the large blue rays of wild succory.

The paths across the corn having no hedges, the wayfarer really walks among the wheat, and can pluck with either hand. The ears rise above the heads of children, who shout with joy as they rush along as though to the arms of their mother.

Beneath the towing-path, at the root of the willow bushes, which the tow-ropes, so often drawn over them, have kept low, the water-docks lift their thick stems and giant leaves. Bunches of rough-leaved comfrey grow down to the water's edge—indeed, the coarse stems sometimes bear signs of having been partially under water when a freshet followed a storm. The flowers are not so perfectly bell-shaped as those of some plants, but are rather tubular. They appear in April, though then green, and may be found all the summer months. Where the comfrey grows thickly the white bells give some colour to the green of the bank, and would give more were they not so often overshadowed by the leaves.

Water betony, or persicaria, lifts its pink spikes everywhere, tiny florets close together round the stem at the top; the leaves are willow-shaped, and there is scarcely a hollow or break in the bank where the earth has fallen which is not clothed with them. A mile or two up the river the tansy is plentiful, bearing golden buttons, which, like every fragment of the feathery foliage, if pressed in the fingers, impart to them a peculiar scent. There, too, the yellow loosestrife pushes up its tall slender stalks to the top of the low willow-bushes, that the bright yellow flowers may emerge from the shadow.

The river itself, the broad stream, ample and full, exhibits all its glory in this reach; from One Tree to the Lock it is nearly straight, and the river itself is everything. Between wooded hills, or where divided by numerous islets, or where trees and hedges enclose the view, the stream is but part of the scene. Here it is all. The long raised bank without a hedge or fence, with the cornfields on its level, simply guides the eye to the water. Those who are afloat upon it insensibly yield to the influence of the open expanse.

The boat whose varnished sides but now slipped so gently that the cutwater did not even raise a wavelet, and every black rivet-head was visible as a line of dots, begins to forge ahead. The oars are dipped farther back, and as the blade feels the water holding it in the hollow, the lissom wood bends to its work. Before the cutwater a wave rises, and, repulsed, rushes outwards. At each stroke, as the weight swings towards the prow, there is just the least faint depression at its stem as the boat travels. Whirlpool after whirlpool glides from the oars, revolving to the rear with a threefold motion, round and round, backwards and outwards. The crew impart their own life to their boat; the animate and inanimate become as one, the boat is no longer wooden but alive.

If there be a breeze a fleet of white sails comes round the willow-hidden bend. But the Thames yachtsmen have no slight difficulties to contend with. The capricious wind is nowhere so thoroughly capricious as on the upper river. Along one mile there may be a spanking breeze, the very next is calm, or with a fitful puff coming over a high hedge, which flutters his pennant, but does not so much as shake the sail. Even in the same mile the wind may take the water on one side, and scarcely move a leaf on the other. But the current is always there, and the vessel is certain to drift.

When at last a good opportunity is obtained, just as the boat heels over, and the rushing bubbles at the prow resound, she must be put about, and the napping foresail almost brushes the osiers. If she does not come round—if the movement has been put off a moment too long—the keel grates, and she is aground immediately. It is nothing but tacking, tacking, tacking—a kind of stitching the stream.

Nor can one always choose the best day for the purpose; the exigencies of business, perhaps, will not permit, and when free, the wind, which has been scattering tiles and chimney-pots and snapping telegraph wires in the City all the week, drops on the Saturday to nothing. He must possess invincible patience, and at the same time be always ready to advance his vessel even a foot, and his judgment must never fail him at the critical time.

But the few brief hours when the circumstances are favourable compensate for delays and monotonous calms; the vessel, built on well-judged lines, answers her helm and responds to his will with instant obedience, and that sense of command is perhaps the great charm of sailing. There are others who find a pleasure in the yacht. When at her moorings on a sunny morning she is sometimes boarded by laughing girls, who have put off from the lawn, and who proceed in the most sailor-like fashion to overhaul the rigging and see that everything is shipshape. No position shows off a well-poised figure to such advantage as when, in a close-fitting costume, a lady's arms are held high above her head to haul at a rope.

So the river life flows by; skiffs, and four oars, canoes, solitary scullers in outriggers, once now and then a swift eight, launches, a bargee in a tublike dingey standing up and pushing his sculls instead of pulling; gentlemen, with their shoulders in a halter, hauling like horses and towing fair freights against the current; and punts poled across to shady nooks. The splashing of oars, the staccato sound as a blade feathered too low meets the wavelets, merry voices, sometimes a song, and always a low undertone, which, as the wind accelerates it, rises to a roar. It is the last eap of the river to the sea; the last weir to whose piles the tide rises. On the bank of the weir where the tide must moisten their roots grow dense masses of willow-herb, almost as high as the shoulder, with trumpet-shaped pink flowers.

Let us go back again to the bank by the cornfields, with the glorious open stretch of stream. In the evening, the rosy or golden hues of the sunset will be reflected on the surface from the clouds; then the bats wheel to and fro, and once now and then a nighthawk will throw himself through the air with uncertain flight, his motions scarcely to be followed, as darkness falls. Am I mistaken, or are kingfishers less numerous than they were only a few seasons since? Then I saw them, now I do not. Long-continued and severe frosts are very fatal to these birds; they die on the perch.

And may I say a word for the Thames otter? The list of really wild animals now existing in the home counties is so very, very short, that the extermination of one of them seems a serious loss. Every effort is made to exterminate the otter. No sooner does one venture down the river than traps, gins, nets, dogs, prongs, brickbats, every species of missile, all the artillery of vulgar destruction, are brought against its devoted head. Unless my memory serves me wrong, one of these creatures caught in a trap not long since was hammered to death with a shovel or a pitchfork.

Now the river fox is, we know, extremely destructive to fish, but what are a basketful of "bait" compared to one otter? The latter will certainly never be numerous, for the moment they become so, otter-hounds would be employed, and then we should see some sport. Londoners, I think, scarcely recognise the fact that the otter is one of the last links between the wild past of ancient England and the present days of high civilisation.

The beaver is gone, but the otter remains, and comes so near the mighty City as just the other side of the well-known Lock, the portal through which a thousand boats at holiday time convey men and women to breathe pure air. The porpoise, and even the seal, it is said, ventures to Westminster sometimes; the otter to Kingston. Thus, the sea sends its denizens past the vast multitude that surges over the City bridges, and the last link with the olden time, the otter, still endeavours to live near.

Perhaps the river is sweetest to look on in spring time or early summer. Seen from a distance the water seems at first sight, when the broad stream fills the vision as a whole, to flow with smooth, even current between meadow and cornfield. But, coming to the brink, that silvery surface now appears exquisitely chased with ever-changing lines. The light airs, wandering to and fro where high banks exclude the direct influence of the breeze, flutter the ripples hither and thither, so that, instead of rolling upon one lee shore, they meet and expend their little force upon each other. A continuous rising and falling, without a line of direction, thus breaks up the light, not with sparkle or glitter, but with endless silvery facets.

There is no pattern. The apparently intertangled tracing on a work of art presently resolves itself into a design, which once seen is always the same. These wavelets form no design; watch the sheeny maze as long as one will, the eye cannot get at the clue, and so unwind the pattern.

Each seems for a second exactly like its fellow, but varies while you say "These two are the same," and the white reflected light upon the wide stream is now strongest here, and instantly afterwards flickers yonder.

Where a gap in the willows admits a current of air a ripple starts to rush straight across, but is met by another returning, which has been repulsed from the bluff bow of a moored boat, and the two cross and run through each other. As the level of the stream now slightly rises and again falls, the jagged top of a large stone by the shore alternately appears above, or is covered by the surface. The water as it retires leaves for a moment a hollow in itself by the stone, and then swings back to fill the vacuum.

Long roots of willows and projecting branches cast their shadow upon the shallow sandy bottom; the shadow of a branch can be traced slanting downwards with the shelve of the sand till lost in the deeper water. Are those little circlets of light enclosing a round umbra or slightly darker spot, that move along the bottom as the bubbles drift above on the surface, shadows or reflections?

In still, dark places of the stream, where there seems no current, a dust gathers on the water, falling from the trees, or borne thither by the wind and dropping where its impulse ceases. Shadows of branches lie here upon the surface itself, received by the greenish water dust. Round the curve on the concave and lee side of the river, where the wind drives the wavelets direct upon the strand, there are little beaches formed by the undermining and fall of the bank.

The tiny surge rolls up the incline; each wave differing in the height to which it reaches, and none of them alike, washing with it minute fragments of stone and gravel, mere specks which vibrate to and fro with the ripple and even drift with the current. Will these fragments, after a process of trituration, ultimately become sand? A groove runs athwart the bottom, left recently by the keel of a skiff, recently only, for in a few hours these specks of gravel, sand, and particles that sweep along the bottom, fill up such depressions. The motion of these atoms is not continuous, but intermittent; now they rise and are carried a few inches and there sink, in a minute or two to rise again and proceed.

Looking to windward there is a dark tint upon the water; but down the stream, turning the other way, intensely brilliant points of light appear and disappear. Behind a boat rowed against the current two widening lines of wavelets, in the shape of an elongated V, stretch apart and glitter, and every dip of the oars and the slippery oar-blades themselves, as they rise out of the water, reflect the sunshine. The boat appears but to touch the surface, instead of sinking into it, for the water is transparent, and the eye can see underneath the keel.

Here, by some decaying piles, a deep eddy whirls slowly round and round; they stand apart from the shore, for the eddy has cleared away the earth around them. Now, walking behind the waves that roll away from you, dark shadowy spots fluctuate to and fro in the trough of the water.

Before a glance can define its shape the shadow elongates itself from a spot to an oval, the oval melts into another oval, and reappears afar off. When, too, in flood time, the hurrying current seems to respond more sensitively to the shape of the shallows and the banks beneath, there boils up from below a ceaseless succession of irregular circles as if the water there expanded from a centre, marking the verge of its outflow with bubbles and raised lines upon the surface.

By the side float tiny whirlpools, some rotating this way and some that, sucking down and boring tubes into the stream. Longer lines wander past, and as they go, curve round, till when about to make a spiral they lengthen out and drift, and thus, perpetually coiling and uncoiling, glide with the current. They somewhat resemble the conventional curved strokes which, upon an Assyrian bas-relief, indicate water.

Under the spring sunshine, the idle stream flows easily onward, yet every part of the apparently even surface varies; and so, too, in a larger way, the aspects of the succeeding reaches change. Upon one broad bend the tints are green, for the river moves softly in a hollow, with its back, as it were, to the wind.

The green lawn sloping to the shore, and the dark cedar's storeys of flattened foliage, tier above tier; the green osiers of two eyots: the light-leaved aspen; the tall elms, fresh and green; and the green hawthorn bushes give their colour to the water, smooth as if polished, in which they are reflected. A white swan floats in the still narrow channel between the eyots, and there is a punt painted green moored in a little inlet by the lawn, and scarce visible under drooping boughs. Roofs of red tile and dormer windows rise behind the trees, the dull yellow of the walls is almost hidden, and deep shadows lurk about the shore.

Opposite, across the stream, a wide green sward stretches beside the towing-path, lit up with sunshine which touches the dandelions till they glow in the grass. From time to time a nightingale sings in a hawthorn unregarded, and in the elms of the park hard by a crowd of jackdaws chatter. But a little way round a curve the whole stream opens to the sunlight and becomes blue, reflecting the sky. Again, sweeping round another curve with bounteous flow, the current meets the wind direct, a cloud comes up, the breeze freshens, and the watery green waves are tipped with foam.

Rolling upon the strand, they leave a line like a tide marked by twigs and fragments of dead wood, leaves, and the hop-like flowers of Chichester elms which have been floated up and left. Over the stormy waters a band of brown bank-martins wheel hastily to and fro, and from the osiers the loud chirp of the sedge-reedling rises above the buffet of the wind against the ear, and the splashing of the waves.

Once more a change, where the stream darts along swiftly, after having escaped from a weir, and still streaked with foam. The shore rises like a sea beach, and on the pebbles men are patching and pitching old barges which have been hauled up on the bank. A skiff partly drawn up on the beach rocks as the current strives to work it loose, and up the varnish of the side glides a flickering light reflected from the wavelets. A fleet of such skiffs are waiting for hire by the bridge; the waterman cleaning them with a parti-coloured mop spies me eyeing his vessels, and before I know exactly what is going on, and whether I have yet made up my mind, the sculls are ready, the cushions in; I take my seat, and am shoved gently forth upon the stream.

After I have gone under the arch, and am clear of all obstructions, I lay the sculls aside, and reclining let the boat drift past a ballast punt moored over the shallowest place, and with a rising load of gravel. One man holds the pole steadying the scoop, while his mate turns a windlass the chain from which drags it along the bottom, filling the bag with pebbles, and finally hauls it to the surface, when the contents are shot out in the punt.

It is a floating box rather than a boat, square at each end, and built for capacity instead of progress. There are others moored in various places, and all hard at work. The men in this one, scarcely glancing at my idle skiff, go steadily on, dropping the scoop, steadying the pole, turning the crank, and emptying the pebbles with a rattle.

Where do these pebbles come from? Like the stream itself there seems a continual supply; if a bank be scooped away and punted to the shore presently another bank forms. If a hollow be deepened, by-and-by it fills up; if a channel be opened, after a while it shallows again. The stony current flows along below, as the liquid current above. Yet in so many centuries the strand has not been cleared of its gravel, nor has it all been washed out from the banks.

The skiff drifts again, at first slowly, till the current takes hold of it and bears it onward. Soon it is evident that a barge-port is near—a haven where barges discharge their cargoes. A by-way leads down to the river where boats are lying for hire—a dozen narrow punts, waiting at this anchorage till groundbait be lawful. The ends of varnished skiffs, high and dry, are visible in a shed carefully covered with canvas; while sheaves of oars and sculls lean against the wooden wall.

Through the open doors of another shed there may be had a glimpse of shavings and tools, and slight battens crossing the workshop in apparent confusion, forming a curious framework. These are the boatbuilder's struts and stays, and contrivances to keep the boat in rigid position, that her lines may be true and delicate, strake upon strake of dull red mahogany rising from the beechen keel, for the craftsman strings his boat almost as a violinist strings his violin, with the greatest care and heed, and with a right adjustment of curve and due proportion. There is not much clinking, or sawing, or thumping; little noise, but much skill.

Gradually the scene opens. Far down a white bridge spans the river; on the shore red-tiled and gabled houses crowd to the very edge; and behind them a church tower stands out clear against the sky. There are barges everywhere. By the towing-path colliers are waiting to be drawn up stream, black as their freight, by the horses that are nibbling the hawthorn hedge; while by the wharf, labourers are wheeling barrows over bending planks from the barges to the carts upon the shore. A tug comes under the bridge, panting, every puff re-echoed from the arches, dragging by sheer force deeply laden flats behind it. The water in front of their bluff bows rises in a wave nearly to the deck, and then swoops in a sweeping curve to the rear.

The current by the port runs back on the wharf side towards its source, and the foam drifts up the river instead of down. Green flags on a sandbank far out in the stream, their roots covered and their bent tips only visible, now swing with the water and now heel over with the breeze. The *Edwin and Angelina* lies at anchor, waiting to be warped into her berth, her sails furled, her green painted water-barrel lashed by the stern, her tiller idle after the long and toilsome voyage from Rochester.

For there are perils of the deep even to those who only go down to it in barges. Barge as she is, she is not without a certain beauty, and a certain interest, inseparable from all that has received the buffet of the salt water, and over which the salt spray has flown. Barge too, as she is, she bears her part in the commerce of the world. The very architecture on the shore is old-fashioned where these bluff-bowed vessels come, narrow streets and overhanging houses, boat anchors in the windows, sails and tarry ropes; and is there not a Row Barge Inn somewhere?

"Hoy, ahoy!"

The sudden shout startles me, and, glancing round, I find an empty black barge, high out of the water, floating helplessly down upon me with the stream. Noiselessly the great hulk had drifted upon me; as it came the light glinted on the wavelets before the bow, quick points of brilliant light. But two strokes with the sculls carried me out of the way.

NUTTY AUTUMN

There is some honeysuckle still flowering at the tops of the hedges, where in the morning gossamer lies like a dewy net. The gossamer is a sign both of approaching autumn and, exactly at the opposite season of the year, of approaching spring. It stretches from pole to pole, and bough to bough, in the copses in February, as the lark sings. It covers the furze, and lies along the hedge-tops in September, as the lark, after a short or partial silence, occasionally sings again.

But the honeysuckle does not flower so finely as the first time; there is more red (the unopened petal) than white, and beneath, lower down the stalk, are the red berries, the fruit of the former bloom. Yellow weed, or ragwort, covers some fields almost as thickly as buttercups in summer, but it lacks the rich colour of the buttercup. Some knotty knapweeds stay in out-of-the-way places, where the scythe has not been; some bunches of mayweed, too, are visible in the corners of the stubble.

Silverweed lays its golden flower—like a buttercup without a stalk—level on the ground; it has no protection, and any passing foot may press it into the dust. A few white or pink flowers appear on the brambles, and in waste places a little St. John's wort remains open, but the seed vessels are for the most part forming. St. John's wort is the flower of the harvest; the yellow petals appear as the wheat ripens, and there are some to be found till the sheaves are carted. Once now and then a blue and slender bell-flower is lighted on; in Sussex the larger varieties bloom till much later.

By still ponds, to which the moorhens have now returned, tall spikes of purple loosestrife rise in bunches. In the furze there is still much

116

yellow, and wherever heath grows it spreads in shimmering gleams of purple between the birches; for these three, furze, heath, and birch are usually together. The fields, therefore, are not yet flowerless, nor yet without colour here and there, and the leaves, which stay on the trees till late in the autumn, are more interesting now than they have been since they lost their first fresh green.

Oak, elm, beech, and birch, all have yellow spots, while retaining their groundwork of green. Oaks are often much browner, but the moisture in the atmosphere keeps the saps in the leaves. Even the birches are only tinted in a few places, the elms very little, and the beeches not much more: so it would seem that their hues will not be gone altogether till November. Frosts have not yet bronzed the dogwood in the hedges, and the hazel leaves are fairly firm. The hazel generally drops its leaves at a touch about this time, and while you are nutting, if you shake a bough, they come down all around.

The rushes are but faintly yellow, and the slender tips still point upwards. Dull purple burrs cover the burdock; the broad limes are withering, but the leaves are thick, and the teazles are still flowering. Looking upwards, the trees are tinted; lower, the hedges are not without colour, and the field itself is speckled with blue and yellow. The stubble is almost hidden in many fields by the growth of weeds brought up by the rain; still the tops appear above and do not allow it to be green. The stubble has a colour—white if barley, yellow if wheat or oats. The meads are as verdant, even more so, than in the spring, because of the rain, and the brooks crowded with green flags.

Haws are very plentiful this year (1881), and exceptionally large, many fully double the size commonly seen. So heavily are the branches laden with bunches of the red fruit that they droop as apple trees do with a more edible burden. Though so big, and to all appearance tempting to birds, none have yet been eaten; and, indeed, haws seem to be resorted to only as a change unless severe weather compels.

Just as we vary our diet, so birds eat haws, and not many of them til driven by frost and snow. If any stay on till the early months of next year wood-pigeons and missel-thrushes will then eat them; but at this season they are untouched. Blackbirds will peck open the hips directly the frost comes; the hips go long before the haws. There was a large crop of mountain-ash berries, every one of which has been taken by blackbirds and thrushes, which are almost as fond of them as of garden fruit.

Blackberries are thick, too—it is a berry year—and up in the horse-chestnut the prickly-coated nuts hang up in bunches, as many as eight in a stalk. Acorns are large, but not so singularly numerous as the berries nor are hazel-nuts. This provision of hedge fruit no more indicates a severe winter than a damaged wheat harvest indicates a mild one.

There is something wrong with elm trees. In the early part of this summer, not long after the leaves were fairly out upon them, here and there a branch appeared as if it had been touched with red-hot iron and burnt up, all the leaves withered and browned on the boughs. First one tree was thus affected, then another, then a third, till, looking round the fields, it seemed as if every fourth or fifth tree had thus been burnt.

It began with the leaves losing colour, much as they do in autumn on the particular bough; gradually they faded, and finally became brown and of course dead. As they did not appear to shrivel up, it looked as if the grub or insect, or whatever did the mischief, had attacked, not the leaves, but the bough itself. Upon mentioning this I found that it had been noticed in elm avenues and groups a hundred miles distant, so that it is not a local circumstance.

As far as yet appears, the elms do not seem materially injured, the damage being outwardly confined to the bough attacked. These brown spots looked very remarkable just after the trees had become green. They were quite distinct from the damage caused by the snow of October 1880. The boughs broken by the snow had leaves upon them which at once turned brown, and in the case of the oak were visible, the following spring, as brown spots among the green. These snapped boughs never bore leaf again. It was the young fresh green leaves of the elms, those that appeared in the spring of 1881, that withered as if scorched. The boughs upon which they grew had not been injured; they were small boughs at the outside of the tree. I hear that this scorching up of elm leaves has been noticed in other districts for several seasons.

The dewdrops of the morning, preserved by the mist, which the sun does not disperse for some hours, linger on late in shaded corners, as under trees, on drooping blades of grass and on the petals of flowers. Wild bees and wasps may often be noticed on these blades of grass that are still wet, as if they could suck some sustenance from the dew. Wasps fight hard for their existence as the nights grow cold. Desperate and ravenous, they will eat anything, but perish by hundreds as the warmth declines.

Dragon-flies of the larger size are now very busy rushing to and fro on their double wings; those who go blackberrying or nutting cannot fail to see them. Only a very few days since—it does not seem a week—there was a chiffchaff calling in a copse as merrily as in the spring. This little bird is the first, or very nearly the first, to come in the spring, and one of the last to go as autumn approaches. It is curious that, though singled out as a first sign of spring, the chiffchaff has never entered into the home life of the people like the robin, the swallow, or even the sparrow.

There is nothing about it in the nursery rhymes or stories, no one goes out to listen to it, children are not taught to recognise it, and grown-up persons are often quite unaware of it. I never once heard a countryman, a labourer, a farmer, or any one who was always out of doors, so much as allude to it. They never noticed it, so much is every one the product of habit.

The first swallow they looked for, and never missed; but they neither heard nor saw the chiffchaff. To those who make any study at all of birds it is, of course, perfectly familiar; but to the bulk of people it is unknown. Yet it is one of the commonest of migratory birds, and sings in every copse and hedgerow, using loud, unmistakable notes. At last, in the middle of September, the chiffchaff, too, is silent. The swallow remains; but for the rest, the birds have flocked together, finches, starlings, sparrows, and gone forth into the midst of the stubble far from the place where their nests were built, and where they sang, and chirped, and whistled so long.

The swallows, too, are not without thought of going. They may be seen twenty in a row, one above the other, or on the slanting ropes or guys which hold up the masts of the rickcloths over the still unfinished corn-ricks. They gather in rows on the ridges of the tiles, and wisely take counsel of each other. Rooks are up at the acorns; they take them from the bough, while the pheasants come underneath and pick up those that have fallen.

The partridge coveys are more numerous and larger than they have been for several seasons, and though shooting has now been practised for more than a fortnight, as many as twelve and seventeen are still to be counted together. They have more cover than usual at this season, not only because the harvest is still about, but because where cut the stubble is so full of weeds that when crouching they are hidden. In some fields the weeds are so thick that even a pheasant can hide.

South of London the harvest commenced in the last week of July. The stubble that was first cut still remains unploughed; it is difficult to find a fresh furrow, and I have only once or twice heard the quick strong puffing of the steam-plough. While the wheat was in shock it was a sight to see the wood-pigeons at it. Flocks of hundreds came perching on the sheaves, and visiting the same field day after day. The sparrows have never had such a feast of grain as this year. Whole corners of wheatfields—they work more at corners—were cleared out as clean by them as if the wheat had been threshed as it stood.

The sunshine of the autumn afternoons is faintly tawny, and the long grass by the wayside takes from it a tawny undertone. Some other colour than the green of each separate blade, if gathered, lies among the bunches, a little, perhaps like the hue of the narrow pointed leaves of he reeds. It is caught only for a moment, and looked at steadily it goes. Among the grass, the hawkweeds, one or two dandelions, and a stray buttercup, all yellow, favour the illusion. By the bushes there is a double row of pale buff bryony leaves; these, too, help to increase the sense of a secondary colour.

The atmosphere holds the beams, and abstracts from them their white brilliance. They come slower with a drowsy light, which casts a less defined shadow of the still oaks. The yellow and brown leaves in the oaks, in the elms, and the beeches, in their turn affect the rays, and retouch them with their own hue. An immaterial mist across the fields looks like a cloud of light hovering on the stubble: the light itself made visible.

The tawniness is indistinct, it haunts the sunshine, and is not to be fixed, any more than you can say where it begins and ends in the complexion of a brunette. Almost too large for their cups, the acorns have a shade of the same hue now before they become brown. As it withers, the many-pointed leaf of the white bryony and the bine as it shrivels, in like manner, do their part. The white thistle-down, which stays on the bursting thistles because there is no wind to waft it away, reflects it; the white is pushed aside by the colour that the stained sunbeams bring.

Pale yellow thatch on the wheat-ricks becomes a deeper yellow; broad roofs of old red tiles smoulder under it. What can you call it but tawniness?—the earth sunburnt once more at harvest time. Sunburnt and brown—for it deepens into brown. Brown partridges, and pheasants, at a distance brown, their long necks stretched in front and long tails behind gleaming in the stubble. Brown thrushes just venturing to sing again. Brown clover hayricks; the bloom on the third crop yonder, which was recently a bright colour, is fast turning brown, too.

Here and there a thin layer of brown leaves rustles under foot. The scaling bark on the lower part of the tree trunks is brown. Dry dock stems, fallen branches, the very shadows, are not black, but brown. With red hips and haws, red bryony and woodbine berries, these together cause the sense rather than the actual existence of a tawny tint. It is pleasant but sunset comes so soon, and then after the trees are in shadow beneath the yellow spots at the tops of the elms still receive the light from the west a few moments longer.

There is something nutty in the short autumn day—shorter than its duration as measured by hours, for the enjoyable day is between the clearing of the mist and the darkening of the shadows. The nuts are ripe and with them is associated wine and fruit. They are hard but tasteful if you eat one, you want ten, and after ten, twenty. In the wine there is a glow, a spot like tawny sunlight; it falls on your hand as you lift the glass.

They are never really nuts unless you gather them yourself. Put down the gun a minute or two, and pull the boughs this way. One or two may drop of themselves as the branch is shaken, one among the brambles another outwards into the stubble. The leaves rustle against hat and shoulders; a thistle is crushed under foot, and the down at last released Bines of bryony hold the ankles, and hazel boughs are stiff and not ready to bend to the will. This large brown nut must be cracked at once; the film slips off the kernel, which is white underneath. It is sweet.

The tinted sunshine comes through between the tall hazel rods; there is a grasshopper calling in the sward on the other side of the mound The bird's nest in the thorn-bush looks as perfect as if just made, instead of having been left long long since—the young birds have flocked into the stubbles. On the briar which holds the jacket the canker rose, which was green in summer, is now rosy. No such nuts as those captured with cunning search from the bough in the tinted sunlight and under the changing leaf.

The autumn itself is nutty, brown, hard, frosty, and sweet. Nuts are hard, frosts are hard; but the one is sweet, and the other braces the strong Exercise often wearies in the spring, and in the summer heats is scarcely to be faced; but in autumn, to those who are well, every step is bracing and hardens the frame, as the sap is hardening in the trees.

ROUND A LONDON COPSE

In October a party of wood-pigeons took up their residence in the little copse which has been previously mentioned. It stands in the angle formed by two suburban roads, and the trees in it overshadow some villa gardens. This copse has always been a favourite with birds, and it is not uncommon to see a pheasant about it, sometimes within gunshot of the gardens, while the call of the partridges in the evening may now and then be heard from the windows. But though frequently visited by wood-pigeons, they did not seem to make any stay till now when this party arrived.

There were eight of them. During the day they made excursions into the stubble fields, and in the evening returned to roost. They remained through the winter, which will be remembered as the most severe for many years. Even in the sharpest frost, if the sun shone out, they called to each other now and then. On the first day of the year their hollow cooing came from the copse at midday.

During the deep snow which blocked the roads and covered the fields almost a foot deep, they were silent, but were constantly observed flying to and fro. Immediately it became milder they recommenced to coo, so that at intervals the note of the wood-pigeon was heard in the adjacent house from October, all through the winter, till the nesting time in May. Sometimes towards sunset in the early spring they all perched together before finally retiring on the bare, slender tips of the tall birch trees, exposed and clearly visible against the sky.

Six once alighted in a row on a long birch branch, bending it down with their weight like a heavy load of fruit. The stormy sunset flamed up, tinting the fields with momentary red, and their hollow voices sounded among the trees. By May they had paired off, and each couple had a part of the copse to themselves. Instead of avoiding the house, they seemed, on the contrary, to come much nearer, and two or three couples built close to the garden.

Just there, the wood being bare of undergrowth, there was nothing to obstruct the sight but some few dead hanging branches, and the pigeons or ringdoves could be seen continually flying up and down from the ground to their nests. They were so near that the darker marking at the end of the tail, as it was spread open to assist the upward flight to the branch, was visible. Outside the garden gate, and not more than twenty yards distant, there stood three young spruce firs, at the edge of the copse, but without the boundary. To the largest of these one of the pigeons came now and then; he was half inclined to choose it for his nest.

The noise of their wings as they rose and threshed their strong feathers together over the tops of the trees was often heard, and while in the garden one might be watched approaching from a distance, swift as the wind, then suddenly half-closing his wings and shooting forwards, he alighted among the boughs. Their coo is not in any sense tuneful; yet it has a pleasant association; for the ringdove is pre-eminently the bird of the woods and forests, and rightly named the wood-pigeon. Yet though so associated with the deepest and most lonely woods, here they were close to the house and garden, constantly heard, and almost always visible; and London, too, so near. They seemed almost as familiar as the sparrows and starlings.

These pigeons were new inhabitants; but turtle-doves had built in the copse since I knew it. They were late coming the last spring I watched them; but, when they did, chose a spot much nearer the house than usual. The turtle-dove has a way of gurgling the soft vowels "oo" in the throat. Swallows do not make a summer, but when the turtle-dove coos summer is certainly come. One afternoon one of the pair flew up into a hornbeam which stood beside the garden not twenty yards at farthest. At first he sat upright on the branch watching me below, then turned and fluttered down to the nest beneath.

While this nesting was going on I could hear five different birds at once either in the garden or from any of the windows. The doves cooed, and every now and then their gentle tones were overpowered by the loud call of the wood-pigeons. A cuckoo called from the top of the tallest birch, and a nightingale and a brook-sparrow (or sedge-reedling) were audible together in the common on the opposite side of the road. It is remarkable that one season there seems more of one kind of bird than the next. The year alluded to, for instance, in this copse was the wood-pigeons' year. But one season previously the copse seemed to belong to the missel-thrushes.

Early in the March mornings I used to wake as the workmen's trains went rumbling by to the great City, to see on the ceiling by the window a streak of sunlight, tinted orange by the vapour through which the level beams had passed. Something in the sense of morning lifts the heart up to the sun. The light, the air, the waving branches speak; the earth and life seem boundless at that moment. In this it is the same on the verge of the artificial City as when the rays come streaming through the pure atmosphere of the Downs. While thus thinking, suddenly there rang out three clear, trumpet-like notes from a tree at the edge of the copse by the garden. A softer song followed, and then again the same three notes, whose wild sweetness echoed through the wood.

The voice of the missel-thrush sounded not only close at hand and in the room, but repeated itself as it floated away, as the bugle-call does. He is the trumpeter of spring: Lord of March, his proud call challenges the woods; there are none who can answer. Listen for the missel-thrush: when he sings the snow may fall, the rain drift, but not for long; the violets are near at hand. The nest was in a birch visible from the garden, and that season seemed to be the missel-thrush's. Another year the cuckoos had possession.

There is a detached ash tree in the field by the copse; it stands apart and about sixty or seventy yards from the garden. A cuckoo came to this ash every morning, and called there for an hour at a time, his notes echoing along the building, one following the other as wavelets roll on the summer sands. After awhile two more used to appear, and then there was a chase round the copse, up to the tallest birch, and out to the ash tree again. This went on day after day, and was repeated every evening. Flying from the ash to the copse and returning, the birds were constantly in sight; they sometimes passed over the house, and the call became so familiar that it was not regarded any more than the chirp of a sparrow. Till the very last the cuckoos remained there, and never ceased to be heard till they left to cross the seas.

That was the cuckoos' season; next spring, they returned again, but much later than usual, and did not call so much, nor were they seen so often while they were there. One was calling in the copse on the evening of the 6th of May as late as half-past eight, while the moon was shining. But they were not so prominent; and as for the missel-thrushes, I did not hear them at all in the copse. It was the wood-pigeons' year. Thus the birds come in succession and reign by turns.

Even the starlings vary, regular as they are by habit. This season (1881) none have whistled on the house-top. In previous years they have always come, and only the preceding spring a pair filled the gutter with the materials of their nest. Long after they had finished a storm descended, and the rain, thus dammed up and unable to escape, flooded the corner. It cost half a sovereign to repair the damage, but it did not matter; the starlings had been happy. It has been a disappointment this year not to listen to their eager whistling and the flutter of their wings as they vibrate them rapidly while hovering a moment before entering their cavern. A pair of house-martins, too, built under the eaves close to the starlings' nest, and they also disappointed me by not returning this season, though the nest was not touched. Some fate, I fear, overtook both starlings and house-martins.

Another time it was the season of the lapwings. Towards the end of November (1881), there appeared a large flock of peewits, or green plovers, which flock passed most of the day in a broad, level ploughed field of great extent. At this time I estimated their number as about four hundred; far exceeding any flock I had previously seen in the neighbourhood. Fresh parties joined the main body continually, until by December there could not have been less than a thousand. Still more and more arrived, and by the first of January (1882) even this number was doubled, and there were certainly fully two thousand there. It is the habit of green plovers to all move at once, to rise from the ground simultaneously, to turn in the air, or to descend—and all so regular that their very wings seem to flap together. The effect of such a vast body of white-breasted birds uprising as one from the dark ploughed earth was very remarkable.

When they passed overhead the air sang like the midsummer hum with the shrill noise of beating wings. When they wheeled a light shot down reflected from their white breasts, so that people involuntarily looked up to see what it could be. The sun shone on them, so that at a distance the flock resembled a cloud brilliantly illuminated. In an instant they turned and the cloud was darkened. Such a great flock had not been seen in that district in the memory of man.

There did not seem any reason for their congregating in this manner, unless it was the mildness of the winter, but winters had been mild before without such a display. The birds as a mass rarely left this one particular field—they voyaged round in the air and settled again in the same place. Some few used to spend hours with the sheep in a meadow, remaining there till dusk, till the mist hid them, and their cry sounded afar in the gloom. They stayed all through the winter, breaking up as the spring approached. By March the great flock had dispersed.

The winter was very mild. There were buttercups, avens, and white nettles in flower on December 31st. On January 7th, there were briar buds opening into young leaf; on the 9th a dandelion in flower, and an arum up. A grey veronica was trying to open flower on the 11th, and hawthorn buds were so far open that the green was visible on the 16th. On February 14th a yellow-hammer sang, and brambles had put forth green buds. Two wasps went by in the sunshine. The 14th is old Candlemas, supposed to rule the weather for some time after. Old Candlemas was very fine and sunny till night, when a little rain fell. The summer that followed was cold and ungenial, with easterly winds, though fortunately it brightened up somewhat for the harvest. A chaffinch sang on the 20th of February: all these are very early dates.

One morning while I was watching these plovers, a man with a gun got over a gate into the road. Another followed, apparently without a weapon, but as the first proceeded to take his gun to pieces, and put the barrel in one pocket at the back of his coat, and the stock in a second, it is possible that there was another gun concealed. The coolness with which the fellow did this on the highway was astounding, but his impudence was surpassed by his stupidity, for at the very moment he hid the gun there was a rabbit out feeding within easy range, which neither of these men observed.

The boughs of a Scotch fir nearly reached to one window. If I recollect rightly, the snow was on the ground in the early part of the year, when a golden-crested wren came to it. He visited it two or three times a week for some time; his golden crest distinctly seen among the dark green needles of the fir.

There are squirrels in the copse, and now and then one comes within sight. In the summer there was one in the boughs of an oak close to the garden. Once, and once only, a pair of them ventured into the garden itself, deftly passing along the wooden palings and exploring a guelder rose-bush. The pheasants which roost in the copse wander to it from distant preserves. One morning in spring, before the corn was up, there was one in a field by the copse calmly walking along the ridge of a furrow so near that the ring round his neck was visible from the road.

In the early part of last autumn, while the acorns were dropping from the oaks and the berries ripe, I twice disturbed a pheasant from the garden of a villa not far distant. There were some oaks hard by, and from under these the bird had wandered into the quiet sequestered garden. The oak in the copse on which the squirrel was last seen is peculiar for bearing oak-apples earlier than any other of the neighbourhood, and there are often half-a-dozen of them on the twigs on the trunk before there is one anywhere else. The famous snowstorm of October 1880 snapped off the leader or top of this oak.

Jays often come, magpies more rarely, to the copse; as for the lesser birds they all visit it. In the hornbeams at the verge blackcaps sing in spring a sweet and cultured song, which does not last many seconds. They visit a thick bunch of ivy in the garden. By these hornbeam trees a streamlet flows out of the copse, crossed at the hedge by a pole, to prevent cattle straying in. The pole is a robin's perch. He is always there, or near; he was there all through the terrible winter, all the summer, and he is there now.

There are a few inches, a narrow strip of sand, beside the streamlet under this pole. Whenever a wagtail dares to come to this sand the robin immediately appears and drives him away. He will bear no intrusion. A pair of butcher-birds built very near this spot one spring, but afterwards appeared to remove to a place where there is more furze, but beside the same hedge. The determination and fierce resolution of the shrike, or butcherbird, despite his small size, is most marked. One day a shrike darted down from a hedge just before me, not a yard in front, and dashed a dandelion to the ground.

His claws clasped the stalk, and the flower was crushed in a moment; he came with such force as to partly lose his balance. His prey was probably a humble-bee which had settled on the dandelion. The shrike's head resembles that of the eagle in miniature. From his favourite branch he surveys the grass, and in an instant pounces on his victim.

There is a quiet lane leading out of one of the roads which have been mentioned down into a wooded hollow, where there are two ponds, one on each side of the lane. Standing here one morning in the early summer, suddenly a kingfisher came shooting straight towards me, and swerving a little passed within three yards; his blue wings, his ruddy front, the white streak beside his neck, and long bill were visible for a moment; then he was away, straight over the meadows, till he cleared a distant hedge and disappeared. He was probably on his way to visit his nest, for though living by the streams kingfishers often have their nest a considerable way from water.

Two years had gone by since I saw one here before, perched then on the trunk of a willow which overhangs one of the ponds. After that came the severe winters, and it seemed as if the kingfishers were killed off, for they are often destroyed by frost, so that the bird came unexpectedly from the shadow of the trees, across the lane, and out into the sunshine over the field. It was a great pleasure to see a kingfisher again.

This hollow is the very place of singing birds in June. Up in the oaks blackbirds whistle—you do not often see them, for they seek the leafy top branches, but once now and then while fluttering across to another perch. The blackbird's whistle is very human, like some one playing the flute; an uncertain player now drawing forth a bar of a beautiful melody and then losing it again. He does not know what quiver or what turn his note will take before it ends; the note leads him and completes itself. His music strives to express his keen appreciation of the loveliness of the days, the golden glory of the meadow, the light, and the luxurious shadows.

Such thoughts can only be expressed in fragments, like a sculptor's chips thrown off as the inspiration seizes him, not mechanically sawn to a set line. Now and again the blackbird feels the beauty of the time, the large white daisy stars, the grass with yellow-dusted tips, the air which comes so softly unperceived by any precedent rustle of the hedge. He feels the beauty of the time, and he must say it. His notes come like wild flowers not sown in order. There is not an oak here in June without a blackbird.

Thrushes sing louder here than anywhere else; they really seem to sing louder, and they are all around. Thrushes appear to vary their notes with the period of the year, singing louder in the summer, and in the mild days of October when the leaves lie brown and buff on the sward under their perch more plaintively and delicately. Warblers and willow-wrens sing in the hollow in June, all out of sight among the trees—they are easily hidden by a leaf.

At that time the ivy leaves which flourish up to the very tops of the oaks are so smooth with enamelled surface, that high up, as the wind moves them, they reflect the sunlight and scintillate. Greenfinches in the elms never cease love-making; and love-making needs much soft talking. A nightingale in a bush sings so loud the hawthorn seems too small for the vigour of the song. He will let you stand at the very verge of the bough; but it is too near, his voice is sweeter across the field.

There are still, in October, a few red apples on the boughs of the trees in a little orchard beside the same road. It is a natural orchard—left to itself—therefore there is always something to see in it. The palings by the road are falling, and are held up chiefly by the brambles about them and the ivy that has climbed up. Trees stand on the right and trees on the left; there is a tall spruce fir at the back.

The apple trees are not set in straight lines: they were at first, but some have died away and left an irregularity; the trees lean this way and that, and they are scarred and marked as it were with lichen and moss. It is the home of birds. A blackbird had its nest this spring in the bushes on the left side, a nightingale another in the bushes on the right, and there the nightingale sang under the shadow of a hornbeam for hours every morning while "City" men were hurrying past to their train.

The sharp relentless shrike that used to live by the copse moved up here, and from that very hornbeam perpetually darted across the road upon insects in the fern and furze opposite. He never entered the orchard; it is often noticed that birds (and beasts of prey) do not touch creatures that build near their own nests. Several thrushes reside in the orchard;

swallows frequently twittered from the tops of the apple trees. As the grass is so safe from intrusion, one of the earliest buttercups flowers here. Bennets—the flower of the grass—come up; the first bennet is to green things what the first swallow is to the breathing creatures of summer.

On a bare bough, but lately scourged by the east wind, the apple bloom appears, set about with the green of the hedges and the dark spruce behind. White horse-chestnut blooms stand up in their stately way, lighting the path which is strewn with the green moss-like flowers fallen from the oaks. There is an early bush of May. When the young apples take form and shape the grass is so high even the buttercups are overtopped by it. Along the edge of the roadside footpath, where the dandelions, plantains, and grasses are thick with seed, the greenfinches come down and feed.

Now the apples are red that are left, and they hang on boughs from which the leaves are blown by every gust. But it does not matter when you pass, summer or autumn, this little orchard has always something to offer. It is not neglected—it is true attention to leave it to itself.

Left to itself, so that the grass reaches its fullest height; so that bryony vines trail over the bushes and stay till the berries fall of their own ripeness; so that the brown leaves lie and are not swept away unless the wind chooses; so that all things follow their own course and bent. The hedge opposite in autumn, when reapers are busy with the sheaves, is white with the large trumpet flowers of the great wild convolvulus (or bindweed). The hedge there seems made of convolvulus then; nothing but convolvulus, and nowhere else does the flower flourish so strongly; the bines remain till the following spring.

Without a path through it, without a border or parterre, unvisited, and left alone, the orchard has acquired an atmosphere of peace and stillness, such as grows up in woods and far-away lonely places. It is so commonplace and unpretentious that passers-by do not notice it; it is merely a corner of meadow dotted with apple trees—a place that needs frequent glances and a dreamy mood to understand it as the birds understand it. They are always there. In spring, thrushes move along, rustling the fallen leaves as they search among the arum sheaths unrolling beside the sheltering palings. There are nooks and corners whence shy creatures can steal out from the shadow and be happy. There is a loving streak of sunshine somewhere among the tree trunks.

Though the copse is so much frequented the migrant birds (which have now for the most part gone) next spring will not be seen nor heard here first. With one exception, it is not the first place to find them. The cuckoos which come to the copse do not call till some time after others have been heard in the neighbourhood. There is another favourite copse a mile distant, and the cuckoo can be heard near it quite a week earlier. This last spring there were two days' difference—a marked interval.

The nightingale that sings in the bushes on the common immediately opposite the copse is late in the same manner. There is a mound about half a mile farther, where a nightingale always sings first, before all the others of the district. The one on the common began to sing last spring a full week later. On the contrary, the sedge-reedling, which chatters side by side with the nightingale, is the first of all his kind to return to the neighbourhood. The same thing happens season after season, so that when once you know these places you can always hear the birds several days before other people.

With flowers it is the same; the lesser celandine, the marsh marigold, the silvery cardamine, appear first in one particular spot, and may be gathered there before a petal has opened elsewhere. The first swallow in this district generally appears round about a pond near some farm buildings. Birds care nothing for appropriate surroundings. Hearing a titlark singing his loudest, I found him perched on the rim of a tub placed for horses to drink from.

This very pond by which the first swallow appears is muddy enough, and surrounded with poached mud, for a herd of cattle drink from and stand in it. An elm overhangs it, and on the lower branches, which are dead, the swallows perch and sing just over the muddy water. A sow lies in the mire. But the sweet swallows sing on softly; they do not see the wallowing animal, the mud, the brown water; they see only the sunshine the golden buttercups, and the blue sky of summer. This is the true way to look at this beautiful earth.

MAGPIE FIELDS

There were ten magpies together on the 9th of September 1881, in a field of clover beside a road but twelve miles from Charing Cross. Ten magpies would be a large number to see at once anywhere in the south, and not a little remarkable so near town. The magpies were doubtless young birds which had packed, and were bred in the nests in the numerous elms of the hedgerows about there. At one time they were scattered over the field, their white and black colours dotted everywhere, so that they seemed to hold entire possession of it.

Then a knot of them gathered together, more came up, and there they were all ten fluttering and restlessly moving. After a while they passed on into the next field, which was stubble, and, collected in a bunch, were even more conspicuous there, as the stubble did not conceal them so much as the clover. That was on the 9th of September; by the end of the month weeds had grown so high that the stubble itself in that field had disappeared, and from a distance it looked like pasture. In the stubble the magpies remained till I could watch them no longer.

A short time afterwards, on the 17th of September, looking over the gateway of an adjacent field which had been wheat, then only recently carried, a pheasant suddenly appeared rising up out of the stubble; and then a second, and a third and fourth. So tall were the weeds that, in a crouching posture, at the first glance they were not visible; then as they fed, stretching their necks out, only the top of their backs could be seen. Presently some more raised their heads in another part of the field, then two more on the left side, and one under an oak by the hedge, till seventeen were counted.

These seventeen pheasants were evidently all young birds, which had wandered from covers, some distance, too, for there is no preserve within a mile at least. Seven or eight came near each other, forming a flock, but just out of gunshot from the road. They were all extremely busy feeding in the stubble. Next day half-a-dozen or so still remained, but the rest had scattered; some had gone across to an acre of barley yet standing in a corner; some had followed the dropping acorns along the hedge into another piece of stubble; others went into a breadth of turnips.

Day by day their numbers diminished as they parted, till only three or four could be seen. Such a sortie from cover is the standing risk of the game-preserver. Towards the end of September, on passing a barley-field, still partly uncut, and with some spread, there was a loud, confused, murmuring sound up in the trees, like that caused by the immense flocks of starlings which collect in winter. The sound, however, did not seem quite the same, and upon investigation it turned out to be an incredible number of sparrows, whose voices were audible across the field.

They presently flew out from the hedge, and alighted on one of the rows of cut barley, making it suddenly brown from one end to the other. There must have been thousands; they continually flew up, swept round with a whirring of wings, and settled, again darkening the spot they chose. Now, as the sparrow eats from morning to night without ceasing, say for about twelve hours, and picks up a grain of corn in the twinkling of an eye, it would be a moderate calculation to allow this vast flock two sacks a week. Among them there was one white sparrow—his white wings showed distinctly among the brown flock. In the most remote country I never observed so great a number of these birds at once; the loss to the farmers must be considerable.

There were a few fine days at the end of the month. One afternoon there rose up a flock of rooks out of a large oak tree standing separate in the midst of an arable field which was then at last being ploughed. This oak is a favourite with the rooks of the neighbourhood, and they have been noticed to visit it more frequently than others. Up they went, perhaps a hundred of them, rooks and jackdaws together cawing and soaring round and round till they reached a great height. At that level, as if they had attained their ballroom, they swept round and round on outstretched wings, describing circles and ovals in the air. Caw-caw! jack-juck-juck! Thus dancing in slow measure, they enjoyed the sunshine, full from their feast of acorns.

Often as one was sailing on another approached and interfered with his course when they wheeled about each other. Soon one dived. Holding his wings at full stretch and rigid, he dived headlong, rotating as he fell, till his beak appeared as if it would be driven into the ground by the violence of the descent. But within twenty feet of the earth he recovered himself and rose again. Most of these dives, for they all seemed to dive in turn, were made over the favourite oak, and they did not rise till they had gone down to its branches. Many appeared about to throw themselves against the boughs.

Whether they wheeled round in circles, or whether they dived, or simply sailed onward in the air, they did it in pairs. As one was sweeping round another came to him. As one sailed straight on a second closely followed. After one had dived the other soon followed, or waited till he had come up and rejoined him. They danced and played in couples as if they were paired already. Some left the main body and steered right away from their friends, but turned and came back, and in about half-an-hour they all descended and settled in the oak from which they had risen. A loud cawing and jack-juck-jucking accompanied this sally.

The same day it could be noticed how the shadows of the elms cast by the bright sunshine on the grass, which is singularly fresh and green this autumn, had a velvety appearance. The dark shadow on the fresh green looked soft as velvet. The waters of the brook had become darker now; they flowed smooth, and at the brink reflected a yellow spray of horse-chestnut. The sunshine made the greenfinches call, the chaffinches utter their notes, and a few thrushes sing; but the latter were soon silenced by frosts in the early morning, which turned the fern to so deep a reddish brown as to approach copper.

At the beginning of October a herd of cows and a small flock of sheep were turned into the clover field to eat off the last crop, the preceding crops having been mown. There were two or more magpies among the sheep every day: magpies, starlings, rooks, crows, and wagtails follow sheep about. The clover this year seems to have been the best crop, though in the district alluded to it has not been without an enemy. Early in July, after the first crop had been mown a short time, there came up a few dull yellowish-looking stalks among it. These increased so much that one field became yellowish all over, the stalks overtopped the clover, and overcame its green.

It was the lesser broom rape, and hardly a clover plant escaped this parasitic growth. By carefully removing the earth with a pocket-knife the two could be dug up together. From the roots of the clover a slender filament passes underground to the somewhat bulbous root of the broom rape, so that although they stand apart and appear separate plants, they are connected under the surface. The stalk of the broom rape is clammy to touch, and is an unwholesome greenish yellow, a dull undecided colour; if cut, it is nearly the same texture throughout. There are numerous dull purplish flowers at the top, but it has no leaves. It is not a pleasant-looking plant—a strange and unusual growth.

One particular field was completely covered with it, and scarcely a clover field in the neighbourhood was perfectly free. But though drawing the sap from the clover plants the latter grew so vigorously that little damage was apparent. After a while the broom rape disappeared, but the clover shot up and afforded good forage. So late as the beginning of October a few poppies flowered in it, their bright scarlet contrasting vividly with the green around, and the foliage above fast turning brown.

The flight of the jay much resembles that of the magpie, the same jaunty, uncertain style, so that at a distance from the flight alone it would be difficult to distinguish them, though in fact the magpie's longer tail and white and black colours always mark him. One morning in July, standing for a moment in the shade beside a birch copse which borders the same road, a jay flew up into the tree immediately overhead, so near that the peculiar shape of the head and bill and all the plumage was visible. He looked down twice, and then flew. Another morning there was a jay on the ground, searching about, not five yards from the road, nor twenty from a row of houses. It was at the corner of a copse which adjoins them. If not so constantly shot at the jay would be anything but wild.

Notwithstanding all these magpies and jays, the partridges are numerous this year in the fields bordering the highway, and which are not watched by keepers. Thinking of the partridges makes me notice the anthills. There were comparatively few this season, but on the 4th of August, which was a sunny day, I saw the inhabitants of a hill beside the road bringing out the eggs into the sunshine. They could not do it fast enough; some ran out with eggs, and placed them on the top of the little mound, and others seized eggs that had been exposed sufficiently and hurried with them into the interior.

Woody nightshade grows in quantities along this road and, apparently, all about the outskirts of the town. There is not a hedge without it, and it creeps over the mounds of earth at the sides of the highways. Some fumitory appeared this summer in a field of barley; till then I had not observed any for some time in that district. This plant, once so common, but now nearly eradicated by culture, has a soft pleasant green. A cornflower, too, flowered in another field, quite a treasure to find where these beautiful blue flowers are so scarce. The last day of August there was a fierce combat on the footpath between a wasp and a brown moth. They rolled over and struggled, now one, now the other uppermost, and the wasp appeared to sting the moth repeatedly. The moth, however, got away.

There are so many jackdaws about the suburbs that, when a flock of rooks passes over, the caw-cawing is quite equalled by the jack-jucking. The daws are easily known by their lesser size and by their flight, for they use their wings three times to the rook's once. Numbers of daws build in the knot-holes and hollows of the horse-chestnut trees in Bushey Park, and in the elms of the grounds of Hampton Court.

To the left of the Diana Fountain there are a number of hawthorn trees, which stand apart, and are aged like those often found on village greens and commons. Upon some of these hawthorns mistletoe grows, not in such quantities as on the apples in Gloucester and Hereford, but in small pieces.

As late in the spring as May-day I have seen some berries, then very large, on the mistletoe here. Earlier in the year, when the adjoining fountain was frozen and crowded with skaters, there were a number of missel-thrushes in these hawthorns, but they appeared to be eating the haws. At all events, they left some of the mistletoe berries, which were on the plant months later.

Just above Molesey Lock, in the meadows beside the towing-path, the blue meadow geranium, or crane's-bill, flowers in large bunches in the summer. It is one of the most beautiful flowers of the field, and after having lost sight of it for some time, to see it again seemed to bring the old familiar far-away fields close to London. Between Hampton Court and Kingston the towing-path of the Thames is bordered by a broad green sward, sufficiently wide to be worth mowing. One July I found a man at work here in advance of the mowers, pulling up yarrow plants with might and main.

The herb grew in such quantities that it was necessary to remove it first, or the hay would be too coarse. On conversing with him, he said that a person came sometimes and took away a trap-load of yarrow; the flowers were to be boiled and mixed with cayenne pepper, as a remedy for cold in the chest. In spring the dandelions here are pulled in sackfuls, to be eaten as salad. These things have fallen so much into disuse in the country that country people are surprised to find the herbalists flourishing round the great city of progress.

The continued dry weather in the early summer of the present year, which was so favourable to partridges and game, was equally favourable to the increase of several other kinds of birds, and among these the jays. Their screeching is often heard in this district, quite as often as it is in country woodlands. One day in the spring I saw six all screeching and yelling together up and down a hedge near the road. Now in October they are plentiful. One flew across overhead with an acorn in its beak, and perched in an elm beside the highway. He pecked at the acorn on the bough, then glanced down, saw me, and fled, dropping the acorn, which fell tap tap from branch to branch till it reached the mound.

Another jay actually flew up into a fir in the green, or lawn, before a farm-house window, crossing the road to do so. Four together were screeching in an elm close to the road, and since then I have seen others with acorns, while walking there. Indeed, this autumn it is not possible to go far without hearing their discordant and unmistakable cry. They were never scarce here, but are unusually numerous this season, and in the scattered trees of hedgerows their ways can be better observed than in the close covert of copses and plantations, where you hear them, but cannot see for the thick fir boughs.

It is curious to note the number of creatures to whom the oak furnishes food. The jays, for instance, are now visiting them for acorns; in the summer they fluttered round the then green branches for the chafers, and in the evenings the fern owls or goat-suckers wheeled about the verge for these and for moths. Rooks come to the oaks in crowds for the acorns; wood-pigeons are even more fond of them, and from their crops quite a handful may sometimes be taken when shot in the trees.

They will carry off at once as many acorns as old-fashioned economical farmers used to walk about with in their pockets, "chucking" them one, two, or three at a time to the pigs in the stye as a *bonne bouche* and an encouragement to fatten well. Never was there such a bird to eat as the wood-pigeon. Pheasants roam out from the preserves after the same fruit, and no arts can retain them at acorn time. Swine are let run out about the

141

hedgerows to help themselves. Mice pick up the acorns that fall, and hide them for winter use, and squirrels select the best.

If there is a decaying bough, or, more particularly, one that has been sawn off, it slowly decays into a hollow, and will remain in that state for years, the resort of endless woodlice, snapped up by insect-eating birds. Down from the branches in spring there descend long, slender threads, like gossamer, with a caterpillar at the end of each—the insect-eating birds decimate these. So that in various ways the oaks give more food to the birds than any other tree. Where there are oaks there are sure to be plenty of birds. Beeches come next. Is it possible that the severe frosts we sometimes have split oak trees? Some may be found split up the trunk and yet not apparently otherwise injured, as they probably would be if it had been done by lightning. Trees are said to burst in America under frost, so that it is not impossible in this country.

There is a young oak beside the highway which in autumn was wreathed as artistically as could have been done by hand. A black bryony plant grew up round it, rising in a spiral. The heart-shaped leaves have dropped from the bine, leaving thick bunches of red and green berries clustering about the greyish stem of the oak.

Every one must have noticed that some trees have a much finer autumn tint than others. This, it will often be found, is an annual occurrence, and the same elm, or beech, or oak that has delighted the eye with its hues this autumn, will do the same next year, and excel its neighbours in colour. Oaks and beeches, perhaps, are the best examples of this, as they are also the trees that present the most beautiful appearance in autumn.

There are oaks on villa lawns near London whose glory of russet foliage in October or November is not to be surpassed in the parks of the country. There are two or three such oaks in Long Ditton. All oaks do not become russet, or buff; some never take those tints. An oak, for instance, not far from those just mentioned never quite loses its green; it cannot be said, indeed, to remain green, but there is a trace of it somewhere; the leaves must, I suppose, be partly buff and partly green; and the mixture

142

of these colours in bright sunshine produces a tint for which I know no accurate term.

In the tops of the poplars, where most exposed, the leaves stay till the last, those growing on the trunk below disappearing long before those on the spire, which bends to every blast. The keys of the hornbeam come twirling down: the hornbeam and the birch are characteristic trees of the London landscape—the latter reaches a great height and never loses its beauty, for when devoid of leaves the feathery spray-like branches only come into view the more.

The abundant bird life is again demonstrated as the evening approaches. Along the hedgerows, at the corners of the copses, wherever there is the least cover, so soon as the sun sinks, the blackbirds announce their presence by their calls. Their "ching-chinging" sounds everywhere; they come out on the projecting branches and cry, then fly fifty yards farther down the hedge, and cry again. During the day they may not have been noticed, scattered as they were under the bushes, but the dusky shadows darkening the fields send them to roost, and before finally retiring, they "ching-ching" to each other.

Then, almost immediately after the sun has gone down, looking to the south-west the sky seen above the trees (which hide the yellow sunset) becomes a delicate violet. Soon a speck of light gleams faintly through it—the merest speck. The first appearance of a star is very beautiful; the actual moment of first contact as it were of the ray with the eye is always a surprise, however often you may have enjoyed it, and notwithstanding that you are aware it will happen. Where there was only the indefinite violet before, the most intense gaze into which could discover nothing, suddenly, as if at that moment born, the point of light arrives.

So glorious is the night that not all London, with its glare and smoke, can smother the sky; in the midst of the gas, and the roar and the driving crowd, look up from the pavement, and there, straight above, are the calm stars. I never forget them, not even in the restless Strand; they face one coming down the hill of the Haymarket; in Trafalgar Square, looking

143

towards the high dark structure of the House at Westminster, the clear bright steel silver of the planet Jupiter shines unwearied, without sparkle or flicker.

Apart from the grand atmospheric changes caused by a storm wave from the Atlantic, or an anti-cyclone, London produces its own sky. Put a shepherd on St. Paul's, allow him three months to get accustomed to the local appearances and the deceptive smoke clouds, and he would then tell what the weather of the day was going to be far more efficiently than the very best instrument ever yet invented. He would not always be right; but he would predict the local London weather with far more accuracy than any one reading the returns from the barometers at Valentia, Stornoway, Brest, or Christiansand.

The reason is this—the barometer foretells the cloud in the sky, but cannot tell where it will burst. The practised eye can judge with very considerable accuracy where the discharge will take place. Some idea of what the local weather of London will be for the next few hours may often be obtained by observation on either of the bridges—Westminster, Waterloo, or London Bridge: there is on the bridges something like a horizon, the best to be got in the City itself, and the changes announce themselves very clearly there. The difference in the definition is really wonderful.

From Waterloo Bridge the golden cross on St. Paul's and the dome at one time stand out as if engraved upon the sky, clear and with a white aspect. At the same time, the brick of the old buildings at the back of the Strand is red and bright. The structures of the bridges appear light, and do not press upon their arches. The yellow straw stacked on the barges is bright, the copper-tinted sails bright, the white wall of the Embankment clear, and the lions' heads distinct. Every trace of colour, in short, is visible.

At another time the dome is murky, the cross tarnished, the outline dim, the red brick dull, the whiteness gone. In summer there is occasionally a bluish haze about the distant buildings. These are the same changes presented by the Downs in the country, and betoken the state of the atmosphere as clearly. The London atmosphere is, I should fancy, quite as well adapted to the artist's uses as the changeless glare of the Continent. The smoke itself is not without its interest.

Sometimes upon Westminster Bridge at night the scene is very striking. Vast rugged columns of vapour rise up behind and over the towers of the House, hanging with threatening aspect; westward the sky is nearly clear, with some relic of the sunset glow: the river itself, black or illuminated with the electric light, imparting a silvery blue tint, crossed again with the red lamps of the steamers. The aurora of dark vapour, streamers extending from the thicker masses, slowly moves and yet does not go away; it is just such a sky as a painter might give to some tremendous historical event, a sky big with presage, gloom, tragedy. How bright and clear, again, are the mornings in summer! I once watched the sun rise on London Bridge, and never forgot it.

In frosty weather, again, when the houses take hard, stern tints, when the sky is clear over great part of its extent, but with heavy thunderous-looking clouds in places—clouds full of snow—the sun becomes of a red or orange hue, and reminds one of the lines of Longfellow when Othere reached the North Cape—

"Round in a fiery ring
Went the great sun, oh King!
With red and lurid light."

The redness of the winter sun in London is, indeed, characteristic.

A sunset in winter or early spring floods the streets with fiery glow. It comes, for instance, down Piccadilly; it is reflected from the smooth varnished roofs of the endless carriages that roll to and fro like the flicker of a mighty fire; it streaks the side of the street with rosiness. The faces of those who are passing are lit up by it, all unconscious as they are. The sky above London, indeed, is as full of interest as above the hills.

Lunar rainbows occasionally occur; two to my knowledge were seen in the direction and apparently over the metropolis recently.

When a few minutes on the rail has carried you outside the hub as it were of London, among the quiet tree-skirted villas, the night reigns as completely as in the solitudes of the country. Perhaps even more so, for the solitude is somehow more apparent. The last theatre-goer has disappeared inside his hall door, the last dull roll of the brougham, with its happy laughing load, has died away—there is not so much as a single footfall. The cropped holly hedges, the leafless birches, the limes and acacias are still and distinct in the moonlight. A few steps farther out on the highway the copse or plantation sleeps in utter silence.

But the tall elms are the most striking; the length of the branches and their height above brings them across the light, so that they stand out even more shapely than when in leaf. The blue sky (not, of course, the blue of day), the white moonlight, the bright stars—larger at midnight and brilliant, in despite of the moon, which cannot overpower them in winter as she does in summer evenings—all are as beautiful as on the distant hills of old. By night, at least, even here, in the still silence, Heaven has her own way.

When the oak leaves first begin to turn buff, and the first acorns drop, the redwings arrive, and their "kuk-kuk" sound in the hedges and the shrubberies in the gardens of suburban villas. They seem to come very early to the neighbourhood of London, and before the time of their appearance in other districts. The note is heard before they are seen; the foliage of the shrubberies, still thick, though changing colour, concealing them. Presently, when the trees are bare, with the exception of a few oaks, they have disappeared, passing on towards the west. The fieldfares too, as I have previously observed, do not stay. But missel-thrushes seem more numerous near town than in the country.

Every mild day in November the thrushes sing; there are meadows where one may be certain to hear the song-thrush. In the dip or valley at Long Ditton there are several meadows well timbered with elm, which

146

are the favourite resorts of thrushes, and their song may be heard just there in the depth of winter, when it would be possible to go a long distance on the higher ground without hearing one. If you hear the note of the song-thrush during frost it is sure to rain within a few hours; it is the first sign of the weather breaking up.

Another autumn sign is the packing (in a sense) of the moorhens. During the summer the numerous brooks and ponds about town are apparently partially deserted by these birds; at least they are not to be seen by casual wayfarers. But directly the winter gets colder they gather together in the old familiar places, and five or six, or even more, come out at once to feed in the meadows or on the lawns by the water.

Green plovers, or peewits, come in small flocks to the fields recently ploughed; sometimes scarcely a gunshot from the walls of the villas. The tiny golden-crested wrens are comparatively numerous near town—the heaths with their bramble thickets doubtless suit them; so soon as the leaves fall they may often be seen.

HERBS

A great green book, whose broad pages are illuminated with flowers, lies open at the feet of Londoners. This volume, without further preface, lies ever open at Kew Gardens, and is most easily accessible from every part of the metropolis. A short walk from Kew station brings the visitor to Cumberland Gate. Resting for a moment upon the first seat that presents itself, it is hard to realise that London has but just been quitted.

Green foliage around, green grass beneath, a pleasant sensation— not silence, but absence of jarring sound—blue sky overhead, streaks and patches of sunshine where the branches admit the rays, wide, cool shadows, and clear, sweet atmosphere. High in a lime tree, hidden from view by the leaves, a chiffchaff sings continually, and from the distance comes the softer note of a thrush. On the close-mown grass a hedge-sparrow is searching about within a few yards, and idle insects float to and fro, visible against the background of a dark yew tree—they could not be seen in the glare of the sunshine. The peace of green things reigns.

It is not necessary to go farther in; this spot at the very entrance is equally calm and still, for there is no margin of partial disturbance— repose begins at the edge. Perhaps it is best to be at once content, and to move no farther; to remain, like the lime tree, in one spot, with the sunshine and the sky, to close the eyes and listen to the thrush. Something, however, urges exploration.

The majority of visitors naturally follow the path, and go round into the general expanse; but I will turn from here sharply to the right, and crossing the sward there is, after a few steps only, another enclosing wall. Within this enclosure, called the Herbaceous Ground, heedlessly passed and perhaps never heard of by the thousands who go to see the Palm Houses, lies to me the real and truest interest of Kew. For here is a living dictionary of English wild flowers.

The meadow and the cornfield, the river, the mountain and the woodland, the seashore, the very waste place by the roadside, each has sent its peculiar representatives, and glancing for the moment, at large, over the beds, noting their number and extent, remembering that the specimens are not in the mass but individual, the first conclusion is that our own country is the true Flowery Land.

But the immediate value of this wonderful garden is in the clue it gives to the most ignorant, enabling any one, no matter how unlearned, to identify the flower that delighted him or her, it may be, years ago, in faraway field or copse. Walking up and down the green paths between the beds, you are sure to come upon it presently, with its scientific name duly attached and its natural order labelled at the end of the patch.

Had I only known of this place in former days, how gladly I would have walked the hundred miles hither! For the old folk, aged men and countrywomen, have for the most part forgotten, if they ever knew, the plants and herbs in the hedges they had frequented from childhood. Some few, of course, they can tell you; but the majority are as unknown to them, except by sight, as, the ferns of New Zealand or the heaths of the Cape. Since books came about, since the railways and science destroyed superstition, the lore of herbs has in great measure decayed and been lost. The names of many of the commonest herbs are quite forgotten—they are weeds, and nothing more. But here these things are preserved; in London, the centre of civilisation and science, is a garden which restores the ancient knowledge of the monks and the witches of the villages.

Thus, on entering to-day, the first plant which I observed is hellebore—a not very common wild herb perhaps, but found in places, and a traditionary use of which is still talked of in the country, a use which I must forbear to mention. What would the sturdy mowers whom

149

I once watched cutting their way steadily through the tall grass in June say
could they see here the black knapweed cultivated as a garden treasure?
Its hard woody head with purple florets lifted high above the ground
was greatly disliked by them, as, too, the blue scabious, and indeed most
other flowers. The stalks of such plants were so much harder to mow
than the grass.

Feathery yarrow sprays, which spring up by the wayside and whereve
the foot of man passes, as at the gateway, are here. White and lilac-tinted
yarrow flowers grow so thickly along the roads round London as ofter
to form a border between the footpath and the bushes of the hedge
Dandelions lift their yellow heads, classified and cultivated—the same
dandelions whose brilliant colour is admired and imitated by artists, and
whose prepared roots are still in use in country places to improve the
flavour of coffee.

Groundsel, despised groundsel—the weed which cumbers the garden
patch, and is hastily destroyed, is here fully recognised. These harebells—
they have flowered a little earlier than in their wild state—how many
scenes they recall to memory! We found them on the tops of the glorious
Downs when the wheat was ripe in the plains and the earth beneath
seemed all golden. Some, too, concealed themselves on the pastures
behind those bunches of tough grass the cattle left untouched. And ever
in cold November, when the mist lifted, while the dewdrops clustered
thickly on the grass, one or two hung their heads under the furze.

Hawkweeds, which many mistake for dandelions; cowslips, in seed
now, and primroses, with foreign primulas around them and enclosed by
small hurdles, foxgloves, some with white and some with red flowers, all
these have their story and are intensely English. Rough-leaved comfrey
of the side of the river and brook, one species of which is so much
talked of as better forage than grass, is here, its bells opening.

Borage, whose leaves float in the claret-cup ladled out to thirsty
travellers at the London railway stations in the hot weather; knotted
figwort, common in ditches; Aaron's rod, found in old gardens; lovely

veronicas; mints and calamints whose leaves, if touched, scent the fingers, and which grow everywhere by cornfield and hedgerow.

This bunch of wild thyme once again calls up a vision of the Downs; it is not so thick and strong, and it lacks that cushion of herbage which so often marks the site of its growth on the noble slopes of the hills, and along the sward-grown fosse of ancient earthworks, but it is wild thyme, and that is enough. From this bed of varieties of thyme there rises up a pleasant odour which attracts the bees. Bees and humble-bees, indeed, buzz everywhere, but they are much too busily occupied to notice you or me.

Is there any difference in the taste of London honey and in that of the country? From the immense quantity of garden flowers about the metropolis it would seem possible for a distinct flavour, not perhaps preferable, to be imparted. Lavender, of which old housewives were so fond, and which is still the best of preservatives, comes next, and self-heal is just coming out in flower; the reapers have, I believe, forgotten its former use in curing the gashes sometimes inflicted by the reap-hook. The reaping-machine has banished such memories from the stubble. Nightshades border on the potato, the flowers of both almost exactly alike; poison and food growing side by side and of the same species.

There are tales still told in the villages of this deadly and enchanted mandragora; the lads sometimes go to the churchyards to search for it. Plantains and docks, wild spurge, hops climbing up a dead fir tree, a well-chosen pole for them—nothing is omitted. Even the silver weed, the dusty-looking foliage which is thrust aside as you walk on the footpath by the road, is here labelled with truth as "cosmopolitan" of habit.

Bird's-foot lotus, another Downside plant, lights up the stones put to represent rockwork with its yellow. Saxifrage, and stone-crop and house-leek are here in variety. Buttercups occupy a whole patch—a little garden to themselves. What would the haymakers say to such a sight? Little, too, does the mower reck of the number, variety, and beauty of the grasses in a single armful of swathe, such as he gathers up to cover his jar of ale with and keep it cool by the hedge. The bennets, the flower of the grass, on their tall stalks, go down in numbers as countless as the sand of the seashore before his scythe.

But here the bennets are watched and tended, the weeds removed from around them, and all the grasses of the field cultivated as affectionately as the finest rose. There is something cool and pleasant in this green after the colours of the herbs in flower, though each grass is but a bunch, yet it has with it something of the sweetness of the meadows by the brooks. Juncus, the rush, is here, a sign often welcome to cattle, for they know that water must be near; the bunch is cut down, and the white pith shows, but it will speedily be up again; horse-tails, too, so thick in marshy places—one small species is abundant in the ploughed fields of Surrey, and must be a great trouble to the farmers, for the land is sometimes quite hidden by it.

In the adjoining water tank are the principal flowers and plants which flourish in brook, river, and pond. This yellow iris flowers in many streams about London, and the water-parsnip's pale green foliage waves at the very bottom, for it will grow with the current right over it as well as at the side. Water-plantain grows in every pond near the metropolis; there is some just outside these gardens, in a wet ha-ha.

The huge water-docks in the centre here flourish at the verge of the adjacent Thames; the marsh marigold, now in seed, blooms in April in the damp furrows of meadows close up to town. But in this flower-pot, sunk so as to be in the water, and yet so that the rim may prevent it from spreading and coating the entire tank with green, is the strangest of all, actually duckweed. The still ponds always found close to cattle yards, are in summer green from end to end with this weed. I recommend all country folk who come up to town in summer time to run down here just to see duckweed cultivated once in their lives.

In front of an ivy-grown museum there is a kind of bowling-green, sunk somewhat below the general surface, where in similar beds may be found the most of those curious old herbs which, for seasoning or salad, or some use of superstition, were famous in ancient English households. Not one of them but has its associations. "There's rue for you," to begin with; we all know who that herb is for ever connected with.

There is marjoram and sage, clary, spearmint, peppermint, salsify, elecampane, tansy, assafoetida, coriander, angelica, caper spurge, lamb's lettuce, and sorrel. Mugwort, southernwood, and wormwood are still to be found in old gardens: they stand here side by side. Monkshood, horehound, henbane, vervain (good against the spells of witches), feverfew, dog's mercury, bistort, woad, and so on, all seem like relics of the days of black-letter books. All the while greenfinches are singing happily in the trees without the wall.

This is but the briefest resume; for many long summer afternoons would be needed even to glance at all the wild flowers that bloom in June. Then you must come once at least a month, from March to September, as the flowers succeed each other, to read the place aright. It is an index to every meadow and cornfield, wood, heath, and river in the country, and by means of the plants of the same species to the flowers of the world. Therefore, the Herbaceous Ground seems to me a place that should on no account be passed by. And the next place is the Wilderness—that is, the Forest.

On the way thither an old-fashioned yew hedge may be seen round about a vast glasshouse. Outside, on the sward, there are fewer wild flowers growing wild than might perhaps be expected, owing in some degree, no doubt, to the frequent mowing, except under the trees, where again the constant shadow does not suit all. By the ponds, in the midst of trees, and near the river, there is a little grass, however, left to itself, in which in June there were some bird's-foot lotus, veronica, hawkweeds, ox-eye daisy, knapweed, and buttercups. Standing by these ponds, I heard a cuckoo call, and saw a rook sail over them; there was no other sound but that of the birds and the merry laugh of children rolling down the slopes.

153

The midsummer hum was audible above; the honey-dew glistened on the leaves of the limes. There is a sense of repose in the mere aspect of large trees in groups and masses of quiet foliage. Their breadth of form steadies the roving eye; the rounded slopes, the wide sweeping outline of these hills of green boughs, induce an inclination, like them, to rest. To recline upon the grass and with half-closed eyes gaze upon them is enough.

The delicious silence is not the silence of night, of lifelessness; it is the lack of jarring, mechanical noise; it is not silence but the sound of leaf and grass gently stroked by the soft and tender touch of the summer air. It is the sound of happy finches, of the slow buzz of humble-bees, of the occasional splash of a fish, or the call of a moorhen. Invisible in the brilliant beams above, vast legions of insects crowd the sky, but the product of their restless motion is a slumberous hum.

These sounds are the real silence; just as a tiny ripple of the water and the swinging of the shadows as the boughs stoop are the real stillness. If they were absent, if it was the soundlessness and stillness of stone, the mind would crave for something. But these fill and content it. Thus reclining, the storm and stress of life dissolve—there is no thought, no care, no desire. Somewhat of the Nirvana of the earth beneath—the earth which for ever produces and receives back again and yet is for ever at rest—enters into and soothes the heart.

The time slips by, a rook emerges from yonder mass of foliage, and idly floats across, and is hidden in another tree. A whitethroat rises from a bush and nervously discourses, gesticulating with wings and tail, for a few moments. But this is not possible for long; the immense magnetism of London, as I have said before, is too near. There comes the quick short beat of a steam launch shooting down the river hard by, and the dream is over. I rise and go on again.

Already one of the willows planted about the pond is showing the yellow leaf, before midsummer. It reminds me of the inevitable autumn. In October these ponds, now apparently deserted, will be full of moorhens.

I have seen and heard but one to-day, but as the autumn comes on they will be here again, feeding about the island, or searching on the sward by the shore. Then, too, among the beeches that lead from hence towards the fanciful pagoda the squirrels will be busy. There are numbers of them, and their motions may be watched with ease. I turn down by the river; in the ditch at the foot of the ha-ha wall is plenty of duckweed, the Lemna of the tank.

A little distance away, and almost on the shore, as it seems, of the Thames, is a really noble horse-chestnut, whose boughs, untouched by cattle, come sweeping down to the ground, and then, continuing, seem to lie on and extend themselves along it, yards beyond their contact. Underneath, it reminds one of sketches of encampments in Hindostan beneath banyan trees, where white tent cloths are stretched from branch to branch. Tent cloths might be stretched here in similar manner, and would enclose a goodly space. Or in the boughs above, a savage's tree-hut might be built, and yet scarcely be seen.

My roaming and uncertain steps next bring me under a plane, and I am forced to admire it; I do not like planes, but this is so straight of trunk, so vast of size, and so immense of height that I cannot choose but look up into it. A jackdaw, perched on an upper bough, makes off as I glance up. But the trees constantly afford unexpected pleasure; you wander among the timber of the world, now under the shadow of the trees which the Red Indian haunts, now by those which grow on Himalayan slopes. The interest lies in the fact that they are trees, not shrubs or mere saplings, but timber trees which cast a broad shadow.

So great is their variety and number that it is not always easy to find an oak or an elm; there are plenty, but they are often lost in the foreign forest. Yet every English shrub and bush is here; the hawthorn, the dogwood, the wayfaring tree, gorse and broom, and here is a round plot of heather. Weary at last, I rest again near the Herbaceous Ground, as the sun declines and the shadows lengthen.

155

As evening draws on, the whistling of blackbirds and the song of thrushes seem to come from everywhere around. The trees are full of them. Every few moments a blackbird passes over, flying at some height, from the villa gardens and the orchards without. The song increases; the mellow whistling is without intermission; but the shadow has nearly reached the wall, and I must go.

TREES ABOUT TOWN

Just outside London there is a circle of fine, large houses, each standing in its own grounds, highly rented, and furnished with every convenience money can supply. If any one will look at the trees and shrubs growing in the grounds about such a house, chosen at random for an example, and make a list of them, he may then go round the entire circumference of Greater London, mile after mile, many days' journey, and find the list ceaselessly repeated.

There are acacias, sumachs, cedar deodaras, araucarias, laurels, planes, beds of rhododendrons, and so on. There are various other foreign shrubs and trees whose names have not become familiar, and then the next grounds contain exactly the same, somewhat differently arranged. Had they all been planted by Act of Parliament, the result could scarcely have been more uniform.

If, again, search were made in these enclosures for English trees and English shrubs, it would be found that none have been introduced. The English trees, timber trees, that are there, grew before the house was built; for the rest, the products of English woods and hedgerows have been carefully excluded. The law is, "Plant planes, laurels, and rhododendrons; root up everything natural to this country."

To those who have any affection for our own woodlands this is a pitiful spectacle, produced, too, by the expenditure of large sums of money. Will no one break through the practice, and try the effect of English trees? There is no lack of them, and they far excel anything yet imported in beauty and grandeur.

Though such suburban grounds mimic the isolation and retiremen of ancient country-houses surrounded with parks, the distinctive featur of the ancient houses is omitted. There are no massed bodies, as it were of our own trees to give a substance to the view. Are young oaks eve seen in those grounds so often described as park-like? Some time sinc it was customary for the builder to carefully cut down every piece o timber on the property before putting in the foundations.

Fortunately, the influence of a better taste now preserves such tree as chance to be growing on the site at the moment it is purchased. Thes remain, but no others are planted. A young oak is not to be seen. Th oaks that are there drop their acorns in vain, for if one takes root it is a once cut off; it would spoil the laurels. It is the same with elms; the ol elms are decaying, and no successors are provided.

As for ash, it is doubtful if a young ash is anywhere to be found; if s it is an accident. The ash is even rarer than the rest. In their places are pu more laurels, cedar deodaras, various evergreens, rhododendrons, planes How tame and insignificant are these compared with the oak! Thrice year the oaks become beautiful in a different way.

In spring the opening buds give the tree a ruddy hue; in summer th great head of green is not to be surpassed; in autumn, with the fallin leaf and acorn, they appear buff and brown. The nobility of the oa casts the pitiful laurel into utter insignificance. With elms it is the same they are reddish with flower and bud very early in the year, the fresh lea is a tender green; in autumn they are sometimes one mass of yellow.

Ashes change from almost black to a light green, then a deeper green and again light green and yellow. Where is the foreign evergreen in th competition? Put side by side, competition is out of the question; yo have only to get an artist to paint the oak in its three phases to see this There is less to be said against the deodara than the rest, as it is a gracefu tree; but it is not English in any sense.

The point, however, is that the foreigners oust the English altogether Let the cedar and the laurel, and the whole host of invading evergreens, b put aside by themselves, in a separate and detached shrubbery, maintained for the purpose of exhibiting strange growths. Let them not crowd the lovely English trees out of the place. Planes are much planted now, with ill effect; the blotches where the bark peels, the leaves which lie on the sward like brown leather, the branches wide apart and giving no shelter to birds—in short, the whole ensemble of the plane is unfit for our country.

It was selected for London plantations, as the Thames Embankment, because its peeling bark was believed to protect it against the deposit of sooty particles, and because it grows quickly. For use in London itself it may be preferable: for semi-country seats, as the modern houses surrounded with their own grounds assume to be, it is unsightly. It has no association. No one has seen a plane in a hedgerow, or a wood, or a copse. There are no fragments of English history clinging to it as there are to the oak.

If trees of the plane class be desirable, sycamores may be planted, as they have in a measure become acclimatised. If trees that grow fast are required, there are limes and horse-chestnuts; the lime will run a race with any tree. The lime, too, has a pale yellow blossom, to which bees resort in numbers, making a pleasant hum, which seems the natural accompaniment of summer sunshine. Its leaves are put forth early.

Horse-chestnuts, too, grow quickly and without any attention, the bloom is familiar, and acknowledged to be fine, and in autumn the large sprays of leaves take orange and even scarlet tints. The plane is not to be mentioned beside either of them. Other trees as well as the plane would have flourished on the Thames Embankment, in consequence of the current of fresh air caused by the river. Imagine the Embankment with double rows of oaks, elms, or beeches; or, if not, even with limes or horse-chestnuts! To these certainly birds would have resorted—possibly rooks, which do not fear cities. On such a site the experiment would have been worth making.

If in the semi-country seats fast-growing trees are needed, there are, as I have observed, the lime and horse-chestnut; and if more variety be desired, add the Spanish chestnut and the walnut. The Spanish chestnut is a very fine tree; the walnut, it is true, grows slowly. If as many beeches as cedar deodaras and laurels and planes were planted in these grounds, in due course of time the tap of the woodpecker would be heard: a sound truly worth ten thousand laurels. At Kew, far closer to town than many of the semi-country seats are now, all our trees flourish in perfection.

Hardy birches, too, will grow in thin soil. Just compare the delicate drooping boughs of birch—they could not have been more delicate if sketched with a pencil—compare these with the gaunt planes!

Of all the foreign shrubs that have been brought to these shores, there is not one that presents us with so beautiful a spectacle as the bloom of the common old English hawthorn in May. The mass of blossom, the pleasant fragrance, its divided and elegant leaf, place it far above any of the importations. Besides which, the traditions and associations of the May give it a human interest.

The hawthorn is a part of natural English life—country life. It stands side by side with the Englishman, as the palm tree is pictured side by side with the Arab. You cannot pick up an old play, or book of the time when old English life was in the prime, without finding some reference to the hawthorn. There is nothing of this in the laurel, or any shrub whatever that may be thrust in with a ticket to tell you its name; it has a ticket because it has no interest, or else you would know it.

For use there is nothing like hawthorn; it will trim into a thick hedge, defending the enclosure from trespassers, and warding off the bitter winds; or it will grow into a tree. Again, the old hedge-crab—the common, despised crab-apple—in spring is covered with blossom, such a mass of blossom that it may be distinguished a mile. Did any one ever see a plane or a laurel look like that?

How pleasant, too, to see the clear white flower of the blackthorn come out in the midst of the bitter easterly breezes! It is like a white handkerchief beckoning to the sun to come. There will not be much more frost; if the wind is bitter to-day, the sun is rapidly gaining power. Probably, if a blackthorn bush were by any chance discovered in the semi-parks or enclosures alluded to, it would at once be rooted out as an accursed thing. The very brambles are superior; there is the flower, the sweet berry, and afterwards the crimson leaves—three things in succession.

What can the world produce equal to the June rose? The common briar, the commonest of all, offers a flower which, whether in itself, or the moment of its appearance at the juncture of all sweet summer things, or its history and associations, is not to be approached by anything a millionaire could purchase. The labourer casually gathers it as he goes to his work in the field, and yet none of the rich families whose names are synonymous with wealth can get anything to equal it if they ransack the earth.

After these, fill every nook and corner with hazel, and make filbert walks. Up and down such walks men strolled with rapiers by their sides while our admirals were hammering at the Spaniards with culverin and demi-cannon, and looked at the sun-dial and adjourned for a game at bowls, wishing that they only had a chance to bowl shot instead of peaceful wood. Fill in the corners with nut-trees, then, and make filbert walks. All these are like old story books, and the old stories are always best.

Still, there are others for variety, as the wild guelder rose, which produces heavy bunches of red berries; dogwood, whose leaves when frost-touched take deep colours; barberry, yielding a pleasantly acid fruit; the wayfaring tree; not even forgetting the elder, but putting it at the outside, because, though flowering, the scent is heavy, and because the elder was believed of old time to possess some of the virtue now attributed to the blue gum, and to neutralise malaria by its own odour.

For colour add the wild broom and some furze. Those who have seen broom in full flower, golden to the tip of every slender bough, cannot need any persuasion, surely, to introduce it. Furze is specked with yellow when the skies are dark and the storms sweep around, besides its prime display. Let wild clematis climb wherever it will. Then laurels may come after these, put somewhere by themselves, with their thick changeless leaves, unpleasant to the touch; no one ever gathers a spray.

Rhododendrons it is unkind to attack, for in themselves they afford a rich flower. It is not the rhododendron, but the abuse of it, which must be protested against. Whether the soil suits or not—and, for the most part, it does not suit—rhododendrons are thrust in everywhere. Just walk in amongst them—behind the show—and look at the spindly, crooked stems, straggling how they may, and then look at the earth under them where not a weed even will grow. The rhododendron is admirable in its place, but it is often overdone and a failure, and has no right to exclude those shrubs that are fitter. Most of the foreign shrubs about these semi-country seats look exactly like the stiff and painted little wooden trees that are sold for children's toys, and, like the toys, are the same colour all the year round.

Now, if you enter a copse in spring the eye is delighted with cowslips on the banks where the sunlight comes, with blue-bells, or earlier with anemones and violets, while later the ferns rise. But enter the semi-parks of the semi-country seat, with its affected assumption of countryness and there is not one of these. The fern is actually purposely eradicated—just think! Purposely! Though indeed they would not grow, one would think, under rhododendrons and laurels, cold-blooded laurels. They will grow under hawthorn, ash, or beside the bramble bushes.

If there chance to be a little pond or "fountain," there is no such thing as a reed, or a flag, or a rush. How the rushes would be hastily hauled out and hurled away with execrations!

Besides the greater beauty of English trees, shrubs, and plants they also attract the birds, without which the grandest plantation is a vacancy, and another interest, too, arises from watching the progress of their growth and the advance of the season. Our own trees and shrubs literally keep pace with the stars which shine in our northern skies. An astronomical floral almanack might almost be constructed, showing how, as the constellations marched on by night, the buds and leaves and flowers appeared by day.

The lower that brilliant Sirius sinks in the western sky after ruling the winter heavens, and the higher that red Arcturus rises, so the buds thicken, open, and bloom. When the Pleiades begin to rise in the early evening, the leaves are turning colour, and the seed vessels of the flowers take the place of the petals. The coincidences of floral and bird life, and of these with the movements of the heavens, impart a sense of breadth to their observation.

It is not only the violet or the anemone, there are the birds coming from immense distances to enjoy the summer with us; there are the stars appearing in succession, so that the most distant of objects seems brought into connection with the nearest, and the world is made one. The sharp distinction, the line artificially drawn between things, quite disappears when they are thus associated.

Birds, as just remarked, are attracted by our own trees and shrubs. Oaks are favourites with rooks and wood-pigeons; blackbirds whistle in them in spring; if there is a pheasant about in autumn he is sure to come under the oak; jays visit them. Elms are resorted to by most of the larger birds. Ash plantations attract wood-pigeons and turtle-doves. Thrushes are fond of the ash, and sing much on its boughs. The beech is the woodpecker's tree so soon as it grows old—birch one of the missel-thrush's.

In blackthorn the long-tailed tit builds the domed nest every one admires. Under the cover of brambles white-throats build. Nightingales love hawthorn, and so does every bird. Plant hawthorn, and almost every bird will come to it, from the wood-pigeon down to the wren. Do not clear away the fallen branches and brown leaves, sweeping the plantation as if it were the floor of a ballroom, for it is just the tangle and the wilderness that brings the birds, and they like the disarray.

If evergreens are wanted, there are the yew, the box, and holly—all three well sanctioned by old custom. Thrushes will come for the yew berries, and birds are fond of building in the thick cover of high box hedges. Notwithstanding the prickly leaves, they slip in and out of the holly easily. A few bunches of rushes and sedges, with some weeds and aquatic grasses, allowed to grow about a pond, will presently bring moorhens. Bare stones—perhaps concrete—will bring nothing.

If a bough falls into the water, let it stay; sparrows will perch on it to drink. If a sandy drinking-place can be made for them the number of birds that will come in the course of the day will be surprising.

Kind-hearted people, when winter is approaching, should have two posts sunk in their grounds, with planks across at the top; a raised platform with the edges projecting beyond the posts, so that cats cannot climb up, and of course higher than a cat can spring. The crumbs cast out upon this platform would gather crowds of birds; they will come to feel at home, and in spring time will return to build and sing.

TO BRIGHTON

The smooth express to Brighton has scarcely, as it seems, left the metropolis when the banks of the railway become coloured with wild flowers. Seen for a moment in swiftly passing, they border the line like a continuous garden. Driven from the fields by plough and hoe, cast out from the pleasure-grounds of modern houses, pulled up and hurled over the wall to wither as accursed things, they have taken refuge on the embankment and the cutting.

There they can flourish and ripen their seeds, little harassed even by the scythe and never by grazing cattle. So it happens that, extremes meeting, the wild flower, with its old-world associations, often grows most freely within a few feet of the wheels of the locomotive. Purple heathbells gleam from shrub-like bunches dotted along the slope; purple knapweeds lower down in the grass; blue scabious, yellow hawkweeds where the soil is thinner, and harebells on the very summit; these are but a few upon which the eye lights while gliding by.

Glossy thistledown, heedless whither it goes, comes in at the open window. Between thickets of broom there is a glimpse down into a meadow shadowed by the trees of a wood. It is bordered with the cool green of brake fern, from which a rabbit has come forth to feed, and a pheasant strolls along with a mind, perhaps, to the barley yonder. Or a foxglove lifts its purple spire; or woodbine crowns the bushes. The sickle has gone over, and the poppies which grew so thick a while ago in the corn no longer glow like a scarlet cloak thrown on the ground. But red spots in waste places and by the ways are where they have escaped the steel.

A wood-pigeon keeps pace with the train—his vigorous pinions can race against an engine, but cannot elude the hawk. He stops presently among the trees. How pleasant it is from the height of the embankment to look down upon the tops of the oaks! The stubbles stretch away, crossed with bands of green roots where the partridges are hiding. Among flags and weeds the moorhens feed fearlessly as we roll over the stream: then comes a cutting, and more heath and hawkweed, harebell, and bramble bushes red with unripe berries.

Flowers grow high up the sides of the quarries; flowers cling to the dry, crumbling chalk of the cliff-like cutting; flowers bloom on the verge above, against the line of the sky, and over the dark arch of the tunnel. This, it is true, is summer; but it is the same in spring. Before a dandelion has shown in the meadow, the banks of the railway are yellow with coltsfoot. After a time the gorse flowers everywhere along them; but the golden broom overtops all, perfect thickets of broom glowing in the sunlight.

Presently the copses are azure with bluebells, among which the brake is thrusting itself up; others, again, are red with ragged robins, and the fields adjacent fill the eye with the gaudy glare of yellow charlock. The note of the cuckoo sounds above the rushing of the train, and the larks may be seen, if not heard, rising high over the wheat. Some birds, indeed find the bushes by the railway the quietest place in which to build their nests.

Butcher-birds or shrikes are frequently found on the telegraph wires from that elevation they pounce down on their prey, and return again to the wire. There were two pairs of shrikes using the telegraph wires for this purpose one spring only a short distance beyond noisy Clapham Junction. Another pair came back several seasons to a particular part of the wires, near a bridge, and I have seen a hawk perched on the wire equally near London.

The haze hangs over the wide, dark plain, which, soon after passing Redhill, stretches away on the right. It seems to us in the train to extend from the foot of a great bluff there to the first rampart of the still distant South Downs. In the evening that haze will be changed to a flood of purple light veiling the horizon. Fitful glances at the newspaper or the novel pass the time; but now I can read no longer, for I know, without any marks or tangible evidence, that the hills are drawing near. There is always hope in the hills.

The dust of London fills the eyes and blurs the vision; but it penetrates deeper than that. There is a dust that chokes the spirit, and it is this that makes the streets so long, the stones so stony, the desk so wooden; the very rustiness of the iron railings about the offices sets the teeth on edge, the sooty blackened walls (yet without shadow) thrust back the sympathies which are ever trying to cling to the inanimate things around us. A breeze comes in at the carriage window—a wild puff, disturbing the heated stillness of the summer day. It is easy to tell where that came from—silently the Downs have stolen into sight.

So easy is the outline of the ridge, so broad and flowing are the slopes, that those who have not mounted them cannot grasp the idea of their real height and steepness. The copse upon the summit yonder looks but a short stroll distant; how much you would be deceived did you attempt to walk thither! The ascent here in front seems nothing, but you must rest before you have reached a third of the way up. Ditchling Beacon there, on the left, is the very highest above the sea of the whole mighty range, but so great is the mass of the hill that the glance does not realise it.

Hope dwells there, somewhere, mayhap, in the breeze, in the sward, or the pale cups of the harebells. Now, having gazed at these, we can lean back on the cushions and wait patiently for the sea. There is nothing else, except the noble sycamores on the left hand just before the train draws into the station.

The clean dry brick pavements are scarcely less crowded than those of London, but as you drive through the town, now and then there is a glimpse of a greenish mist afar off between the houses. The green mist thickens in one spot almost at the horizon; or is it the dark nebulous sails

167

of a vessel? Then the foam suddenly appears close at hand—a white streak seems to run from house to house, reflecting the sunlight: and this is Brighton.

"How different the sea looks away from the pier!" It is a new pleasure to those who have been full of gaiety to see, for once, the sea itself. Westwards, a mile beyond Hove, beyond the coastguard cottages, turn aside from the road, and go up on the rough path along the ridge of shingle. The hills are away on the right, the sea on the left; the yards of the ships in the basin slant across the sky in front.

With a quick, sudden heave the summer sea, calm and gleaming, runs a little way up the side of the groyne, and again retires. There is scarce a gurgle or a bubble, but the solid timbers are polished and smooth where the storms have worn them with pebbles. From a grassy spot ahead a bird rises, marked with white, and another follows it; they are wheatears; they frequent the land by the low beach in the autumn.

A shrill but feeble pipe is the cry of the sandpiper, disturbed on his moist feeding-ground. Among the stones by the waste places there are pale-green wrinkled leaves, and the large yellow petals of the sea-poppy. The bright colour is pleasant, but it is a flower best left ungathered, for its odour is not sweet. On the wiry sward the light pink of the sea-daisies (or thrift) is dotted here and there: of these gather as you will. The presence even of such simple flowers, of such well-known birds, distinguishes the solitary from the trodden beach. The pier is in view, but the sea is different here.

Drive eastwards along the cliffs to the rough steps cut down to the beach, descend to the shingle, and stroll along the shore to Rottingdean. The buttresses of chalk shut out the town if you go to them, and rest near the large pebbles heaped at the foot. There is nothing but the white cliff, the green sea, the sky, and the slow ships that scarcely stir.

In the spring, a starling comes to his nest in a cleft of the cliff above; he shoots over from the dizzy edge, spreads his wings, borne up by the ascending air, and in an instant is landed in his cave. On the sward above, n the autumn, the yellow lip of the toad-flax, spotted with orange, peers from the grass as you rest and gaze—how far?—out upon the glorious plain.

Or go up on the hill by the race-course, the highest part near the sea, and sit down there on the turf. If the west or south wind blow ever so slightly the low roar of the surge floats up, mingling with the rustle of the corn stacked in shocks on the slope. There inhale unrestrained the breeze, the sunlight, and the subtle essence which emanates from the ocean. For the loneliest of places are on the borders of a gay crowd, and thus in Brighton—the by-name for all that is crowded and London-like—it is possible to dream on the sward and on the shore.

In the midst, too, of this most modern of cities, with its swift, luxurious service of Pullman cars, its piers, and social pleasures, there exists a collection which, in a few strokes, as it were, sketches the ways and habits and thoughts of old rural England. It is not easy to realise in these days of quick transit and still quicker communication that old England was mostly rural.

There were towns, of course, seventy years ago, but even the towns were penetrated with what, for want of a better word, may be called country sentiment. Just the reverse is now the case; the most distant hamlet which the wanderer in his autumn ramblings may visit, is now more or less permeated with the feelings and sentiment of the city. No written history has preserved the daily life of the men who ploughed the Weald behind the hills there, or tended the sheep on the Downs, before our beautiful land was crossed with iron roads; while news, even from the field of Waterloo, had to travel slowly. And, after all, written history is but words, and words are not tangible.

But in this collection of old English jugs, and mugs, and bowls, and cups, and so forth, exhibited in the Museum, there is the real presentment of old rural England. Feeble pottery has ever borne the impress of man more vividly than marble. From these they quenched their thirst, over these they laughed and joked, and gossiped, and sang old hunting songs

till the rafters rang, and the dogs under the table got up and barked.
Cannot you see them? The stubbles are ready now once more for the
sportsmen.

With long-barrelled flint-lock guns they ranged over that wonderful
map of the land which lies spread out at your feet as you look down from
the Dyke. There are already yellowing leaves; they will be brown after a
while, and the covers will be ready once more for the visit of the hounds.
The toast upon this mug would be very gladly drunk by the agriculturist
of to-day in his silk hat and black coat. It is just what he has been wishing
these many seasons.

> "Here's to thee, mine honest friend,
> Wishing these hard times to mend."

Hard times, then, are nothing new.

"It is good ale," is the inscription on another jug; that jug would be
very welcome if so filled in many a field this very day. "Better luck still" is
a jug motto which every one who reads it will secretly respond to. Cock-
fighting has gone by, but we are even more than ever on the side of fair
play, and in that sense can endorse the motto, "May the best cock win."
A cup desires that fate should give

> "Money to him who has spirit to use it,
> And life to him who has courage to lose it."

A mug is moderate of wishes and somewhat cynical:—

> "A little health, a little wealth,
>   A little house, and freedom;
> And at the end a little friend,
>   And little cause to need him."

The toper, if he drank too deep, sometimes found a frog or newt at the bottom (in china)—a hint not to be too greedy. There seem to have been sad dogs about in those days from the picture on this piece—one sniffing regretfully at the bunghole of an empty barrel:—

"This cask when stored with gin I loved to taste,
But now a smell, alas! must break my fast."

Upon a cup a somewhat Chinese arrangement of words is found:—

More	beer	score	Clarke
for	my	the	his
do	trust	pay	sent
I	I	must	has
shall	if	you	maltster
what	for	and	the

These parallel columns can be deciphered by beginning at the last word, "the," on the right hand, and reading up. With rude and sometimes grim humour our forefathers seem to have been delighted. The teapots of our great grandmothers are even more amusingly inscribed and illustrated. At Gretna Green the blacksmith is performing a "Red-Hot Marriage," using his anvil for the altar.

"Oh! Mr. Blacksmith, ease our pains,
And tie us fast in wedlock's chains."

The china decorated with vessels and alluding to naval matters shows how popular was the navy, and how deeply everything concerning Nelson's men had sunk into the minds of the people. Some of the line of battleships here represented are most cleverly executed—every sail and rope and gun brought out with a clearness which the best draughtsman

171

could hardly excel. It is a little hard, however, to preserve the time-honoured imputation upon Jack's constancy in this way on a jug:—

"A sailor's life's a pleasant life,
   He freely roams from shore to shore;
In every port he finds a wife—
   What can a sailor wish for more?"

Some enamoured potter having produced a masterpiece as a present to his lady destroyed the design, so that the service he gave her might be unique. After gazing at these curious old pieces, with dates of 1754, 1728, and so forth, the mind becomes attuned to such times, and the jug with the inscription, "Claret, 1652," seems quite an easy and natural transition.

From the Brighton of to-day it is centuries back to 1754; but from 1754 to 1652 is but a year or two. And after studying these shelves, and getting, as it were, so deep down in the past, it is with a kind of Rip Van Winkle feeling that you enter again into the sunshine of the day. The fair upon the beach does not seem quite real for a few minutes.

Before the autumn is too far advanced and the skies are uncertain, a few hours should be given to that massive Down which fronts the traveller from London, Ditchling Beacon, the highest above the sea-level. It is easy of access, the train carries you to Hassock's Gate—the station is almost in a copse—and an omnibus runs from it to a comfortable inn in the centre of Ditchling village. Thence to the Down itself the road is straight and the walk no longer than is always welcome after riding.

After leaving the cottages and gardens, the road soon becomes enclosed with hedges and trees, a mere country lane; and how pleasant are the trees after the bare shore and barren sea! The hand of autumn has browned the oaks, and has passed over the hedge, reddening the haws. The north wind rustles the dry hollow stalks of plants upon the mound, and there is a sense of hardihood in the touch of its breath.

The light is brown, for a vapour conceals the sun—it is not like a cloud, for it has no end or outline, and it is high above where the summer blue was lately. Or is it the buff leaves, the grey stalks, the dun grasses, the ripe fruit, the mist which hides the distance that makes the day so brown? But the ditches below are yet green with brooklime and rushes. By a gateway stands a tall campanula or bell-flower, two feet high or nearly, with great bells of blue.

A passing shepherd, without his sheep, but walking with his crook as a staff, stays and turns a brown face towards me when I ask him the way. He points with his iron crook at a narrow line which winds up the Down by some chalk-pits; it is a footpath from the corner of the road. Just by the corner the hedge is grey with silky flocks of clematis; the hawthorn is hidden by it. Near by there is a bush, made up of branches from five different shrubs and plants.

First hazel, from which the yellow leaves are fast dropping; among this dogwood, with leaves darkening; between these a bramble bearing berries, some red and some ripe, and yet a pink flower or two left. Thrusting itself into the tangle, long woody bines of bittersweet hang their clusters of red berries, and above and over all the hoary clematis spreads its beard, whitening to meet the winter. These five are all intermixed and bound up together, flourishing in a mass; nuts and edible berries, semi-poisonous fruit, flowers, creepers; and hazel, with markings under its outer bark like a gun-barrel.

This is the last of the plain. Now every step exposes the climber to the force of the unchecked wind. The harebells swing before it, the bennets whistle, but the sward springs to the foot, and the heart grows lighter as the height increases. The ancient hill is alone with the wind. The broad summit is left to scattered furze and fern cowering under its shelter. A sunken fosse and earthwork have slipped together. So lowly are they now after these fourteen hundred years that in places the long rough grass covers and conceals them altogether.

Down in the hollow the breeze does not come, and the bennets do not whistle, yet gazing upwards at the vapour in the sky I fancy I can hear the mass, as it were, of the wind going over. Standing presently at the edge of the steep descent looking into the Weald, it seems as if the mighty blast rising from that vast plain and glancing up the slope like an arrow from a tree could lift me up and bear me as it bears a hawk with outspread wings.

A mist which does not roll along or move is drawn across the immense stage below like a curtain. There is indeed, a brown wood beneath; but nothing more is visible. The plain is the vaster for its vague uncertainty. From the north comes down the wind, out of the brown autumn light, from the woods below and twenty miles of stubble. Its stratum and current is eight hundred feet deep.

Against my chest, coming up from the plough down there (the old plough, with the shaft moving on a framework with wheels), it hurls itself against the green ramparts, and bounds up savagely at delay. The ears are filled with a continuous sense of something rushing past; the shoulders go back square; an iron-like feeling enters into the sinews. The air goes through my coat as if it were gauze, and strokes the skin like a brush.

The tide of the wind, like the tide of the sea, swirls about, and its cold push at the first causes a lifting feeling in the chest—a gulp and pant—as if it were too keen and strong to be borne. Then the blood meets it, and every fibre and nerve is filled with new vigour. I cannot drink enough of it. This is the north wind.

High as is the hill, there are larks yonder singing higher still, suspended in the brown light. Turning away at last and tracing the fosse, there is at the point where it is deepest and where there is some trifling shelter, a flat hawthorn bush. It has grown as flat as a hurdle, as if trained espalierwise or against a wall—the effect, no doubt, of the winds. Into and between its gnarled branches, dry and leafless, furze boughs have been woven in and out, so as to form a shield against the breeze. On the lee of this natural hurdle there are black charcoal fragments and ashes, where a fire has burnt itself out; the stick still leans over on which was hung the vessel used at this wild bivouac.

Descending again by the footpath, the spur of the hill yonder looks larger and steeper and more ponderous in the mist; it seems higher than this, a not unusual appearance when the difference in altitude is not very great. The level we are on seems to us beneath the level in the distance, as the future is higher than the present. In the hedge or scattered bushes, half-way down by the chalk-pit, there grows a spreading shrub—the wayfaring tree—bearing large, broad, downy leaves and clusters of berries, some red and some black, flattened at their sides. There are nuts, too, here, and large sloes or wild bullace. This Ditchling Beacon is, I think, the nearest and the most accessible of the southern Alps from London; it is so near it may almost be said to be in the environs of the capital. But it is alone with the wind.

THE SOUTHDOWN SHEPHERD

The shepherd came down the hill carrying his greatcoat slung at his back upon his crook, and balanced by the long handle projecting in front. He was very ready and pleased to show his crook, which, however, was not so symmetrical in shape as those which are represented upon canvas. Nor was the handle straight; it was a rough stick—the first, evidently, that had come to hand.

As there were no hedges or copses near his walks, he had to be content with this bent wand till he could get a better. The iron crook itself he said was made by a blacksmith in a village below. A good crook was often made from the barrel of an old single-barrel gun, such as in their decadence are turned over to the bird-keepers.

About a foot of the barrel being sawn off at the muzzle end, there was a tube at once to fit the staff into, while the crook was formed by hammering the tough metal into a curve upon the anvil. So the gun—the very symbol of destruction—was beaten into the pastoral crook, the emblem and implement of peace. These crooks of village workmanship are now subject to competition from the numbers offered for sale at the shops at the market towns, where scores of them are hung up on show, all exactly alike, made to pattern, as if stamped out by machinery.

Each village-made crook had an individuality, that of the blacksmith—somewhat rude, perhaps, but distinctive—the hand shown in the iron. While talking, a wheatear flew past, and alighted near the path—a place they frequent. The opinion seems general that wheatears are not so numerous as they used to be. You can always see two or three on the

Downs in autumn, but the shepherd said years ago he had heard of one man catching seventy dozen in a day.

Perhaps such wholesale catches were the cause of the comparative deficiency at the present day, not only by actual diminution of numbers, but in partially diverting the stream of migration. Tradition is very strong in birds (and all animated creatures); they return annually in the face of terrible destruction, and the individuals do not seem to comprehend the danger. But by degrees the race at large becomes aware of and acknowledges the mistake, and slowly the original tracks are deserted. This is the case with water-fowl, and even, some think, with sea-fish.

There was not so much game on the part of the hills he frequented as he had known when he was young, and with the decrease of the game the foxes had become less numerous. There was less cover as the furze was ploughed up. It paid, of course, better to plough it up, and as much as an additional two hundred acres on a single farm had been brought under the plough in his time. Partridges had much decreased, but there were still plenty of hares: he had known the harriers sometimes kill two dozen a day.

Plenty of rabbits still remained in places. The foxes' earths were in their burrows or sometimes under a hollow tree, and when the word was sent round the shepherds stopped them for the hunt very early in the morning. Foxes used to be almost thick. He had seen as many as six (doubtless the vixen and cubs) sunning themselves on the cliffs at Beachy Head, lying on ledges before their inaccessible breeding-places, in the face of the chalk.

At present he did not think there were more than two there. They ascended and descended the cliff with ease, though not, of course, the straight wall or precipice. He had known them fall over and be dashed to pieces, as when fighting on the edge, or in winter by the snow giving way under them. As the snow came drifting along the summit of the Down it gradually formed a projecting eave or cornice, projecting the length of the arm, and frozen.

Something like this may occasionally be seen on houses when the partially melted snow has frozen again before it could quite slide off. Walking on this at night, when the whole ground was white with snow and no part could be distinguished, the weight of the fox as he passed a weak place caused it to give way, and he could not save himself. Last winter he had had two lambs, each a month old, killed by a fox which ate the heads and left the bodies; the fox always eating the head first, severing it, whether of a hare, rabbit, duck, or the tender lamb, and "covering"— digging a hole and burying—that which he cannot finish. To the buried carcase the fox returns the next night before he kills again.

His dog was a cross with a collie: the old sheep-dogs were shaggier and darker. Most of the sheep-dogs now used were crossed with the collie, either with Scotch or French, and were very fast—too fast in some respects. He was careful not to send them much after the flock, especially after feeding, when, in his own words, the sheep had "best walk slow then, like folk"—like human beings, who are not to be hastened after a meal. If he wished his dog to fetch the flock, he pointed his arm in the direction he wished the dog to go, and said, "Put her back." Often it was to keep the sheep out of turnips or wheat, there being no fences. But he made it a practice to walk himself on the side where care was needed, so as not to employ the dog unless necessary.

There is something almost Australian in the wide expanse of South Down sheepwalks, and in the number of the flocks, to those who have been accustomed to the small sheltered meadows of the vales, where forty or fifty sheep are about the extent of the stock on many farms. The land, too, is rented at colonial prices, but a few shillings per acre, so different from the heavy meadow rents. But, then, the sheep-farmer has to occupy a certain proportion of arable land as well as pasture, and here his heavy losses mainly occur.

There is nothing, in fact, in this country so carefully provided against as the possibility of an English farmer becoming wealthy. Much downland is covered with furze; some seems to produce a grass too coarse, so that the rent is really proportional. A sheep to an acre is roughly the allowance.

From all directions along the roads the bleating flocks concentrate at the right time upon the hillside where the sheep-fair is held. You can go nowhere in the adjacent town except uphill, and it needs no hand-post to the fair to those who know a farmer when they see him, the stream of folk tender thither so plainly. It rains, as the shepherd said it would; the houses keep off the drift somewhat in the town, but when this shelter is left behind, the sward of the hilltop seems among the clouds.

The descending vapours close in the view on every side. The actual field underfoot, the actual site of the fair, is visible, but the surrounding valleys and the Downs beyond them are hidden with vast masses of grey mist. For a moment, perhaps, a portion may lift as the breeze drives it along, and the bold, sweeping curves of a distant hill appear, but immediately the rain falls again and the outline vanishes. The glance can only penetrate a few hundred yards; all beyond that becomes indistinct, and some cattle standing higher up the hill are vague and shadowy.

Like a dew, the thin rain deposits a layer of tiny globules on the coat; the grass is white with them hurdles, flakes, everything is as it were the eighth of an inch deep in water. Thus on the hillside, surrounded by the clouds, the fair seems isolated and afar off. A great cart-horse is being trotted out before the little street of booths to make him show his paces; they flourish the first thing at hand—a pole with a red flag at the end—and the huge frightened animal plunges hither and thither in clumsy terror. You must look out for yourself and keep an eye over your shoulder, except among the sheep-pens.

There are thousands of sheep, all standing with their heads uphill. At the corner of each pen the shepherd plants his crook upright: some of them have long brown handles, and these are of hazel with the bark on; others are ash, and one of willow. At the corners, too, just outside, the dogs are chained, and, in addition, there is a whole row of dogs fastened to the tent pegs. The majority of the dogs thus collected together from many miles of the Downs are either collies, or show a very decided trace of the collie.

One old shepherd, an ancient of the ancients, grey and bent, has spent so many years among his sheep that he has lost all notice and observation—there is no "speculation in his eye" for anything but his sheep. In his blue smock frock, with his brown umbrella, which he has had no time or thought to open, he stands listening, all intent, to the conversation of the gentlemen who are examining his pens. He leads a young restless collie by a chain; the links are polished to a silvery brightness by continual motion; the collie cannot keep still; now he runs one side, now the other, bumping the old man, who is unconscious of everything but the sheep.

At the verge of the pens there stand four oxen with their yokes, and the long slender guiding-rod of hazel placed lightly across the necks of the two foremost. They are quite motionless, except their eyes, and the slender rod, so lightly laid across, will remain without falling. After traversing the whole field, if you return you will find them exactly in the same position. Some black cattle are scattered about on the high ground in the mist, which thickens beyond them, and fills up the immense hollow of the valley.

In the street of booths there are the roundabouts, the swings, the rifle galleries—like shooting into the mouth of a great trumpet—the shows, the cakes and brown nuts and gingerbread, the ale-barrels in a row, the rude forms and trestle tables; just the same, the very same, we saw at our first fair five-and-twenty years ago, and a hundred miles away. It is just the same this year as last, like the ploughs and hurdles, and the sheep themselves. There is nothing new to tempt the ploughboy's pennies— nothing fresh to stare at.

The same thing year after year, and the same sounds—the dismal barrel organs, and brazen instruments, and pipes, wailing, droning, booming. How melancholy the inexpressible noise when the fair is left behind, and the wet vapours are settling and thickening around it! But the melancholy s not in the fair—the ploughboy likes it; it is in ourselves, in the thought that thus, though the years go by, so much of human life remains the same—the same blatant discord, the same monotonous roundabout, the same poor gingerbread.

The ploughs are at work, travelling slowly at the ox's pace up and down the hillside. The South Down plough could scarcely have been invented; it must have been put together bit by bit in the slow years—slower than the ox; it is the completed structure of long experience. It is made of many pieces, chiefly wood, fitted and shaped and worked, as it were, together, well seasoned first, built up, like a ship, by cunning of hand.

None of these were struck out—a hundred a minute—by irresistible machinery ponderously impressing its will on iron as a seal on wax—a hundred a minute, and all exactly alike. These separate pieces which compose the plough were cut, chosen, and shaped in the wheelwright's workshop, chosen by the eye, guided in its turn by long knowledge of wood, and shaped by the living though hardened hand of man. So complicated a structure could no more have been struck out on paper in a deliberate and single plan than those separate pieces could have been produced by a single blow.

There are no machine lines—no lines filed out in iron or cut by the lathe to the draughtsman's design, drawn with straight-edge and ruler on paper. The thing has been put together bit by bit: how many thousand, thousand clods must have been turned in the furrows before the idea arose, and the curve to be given to this or that part grew upon the mind as the branch grows on the tree! There is not a sharp edge or sharp corner in it; it is all bevelled and smoothed and fluted as if it had been patiently carved with a knife, so that, touch it where you will, it handles pleasantly.

In these curved lines and smoothness, in this perfect adaptability of means to end, there is the spirit of art showing itself, not with colour or crayon, but working in tangible material substance. The makers of this plough—not the designer—the various makers, who gradually put it together, had many things to consider. The fields where it had to work were, for the most part, on a slope, often thickly strewn with stones which jar and fracture iron.

The soil was thin, scarce enough on the upper part to turn a furrow deepening to nine inches or so at the bottom. So quickly does the rain sink in, and so quickly does it dry, that the teams work in almost every weather, while those in the vale are enforced to idleness. Drain furrows were not needed, nor was it desirable that the ground should be thrown up in "lands," rising in the centre. Oxen were the draught animals, patient enough, but certainly not nimble. The share had to be set for various depths of soil.

All these are met by the wheel plough, and in addition it fulfils the indefinite and indefinable condition of handiness. A machine may be apparently perfect, a boat may seem on paper, and examined on principles the precise build, and yet when the one is set to work and the other floated they may fail. But the wheel plough, having grown up, as it were out of the soil, fulfils the condition of handiness.

This handiness, in fact, embraces a number of minor conditions which can scarcely be reduced to writing, but which constantly occur in practice, and by which the component parts of the plough were doubtless unconsciously suggested to the makers. Each has its proper name. The framework, on wheels in front—the distinctive characteristic of the plough—is called collectively "tacks," and the shafts of the plough rest on it loosely, so that they swing or work almost independently, not unlike a field-gun limbered up.

The pillars of the framework have numerous holes, so that the plough can be raised or lowered, that the share may dig deep or shallow. Then there is the "cock-pin," the "road-bat" (a crooked piece of wood), the "sherve-wright" (so pronounced)—shelvewright (?)—the "rist," and spindle, besides, of course, the usual coulter and share. When the oxen arrive at the top of the field, and the first furrow is completed, they stop well knowing their duty, while the ploughman moves the iron rist, and the spindle which keeps it in position, to the other side, and moves the road-bat so as to push the coulter aside. These operations are done in a minute, and correspond in some degree to turning the rudder of a ship. The object is that the plough, which has been turning the earth one way, shall now (as it is reversed to go downhill) continue to turn it that way. If the change were not effected when the plough was swung round, the furrow would be made opposite. Next he leans heavily on the handles, still standing on the same spot; this lifts the plough, so that it turns easily as if on a pivot.

Then the oxen "jack round"—that is, walk round—so as to face downhill, the framework in front turning like the fore-wheels of a carriage. So soon as they face downhill and the plough is turned, they commence work and make the second furrow side by side with the first. The same operation is repeated at the bottom, and thus the plough travels straight up and down, always turning the furrow the same way, instead of, as in the valleys, making a short circuit at each end, and throwing the earth in opposite directions. The result is a perfectly level field, which, though not designed for it, must suit the reaping-machine better than the drain furrows and raised "lands" of the valley system.

It is somewhat curious that the steam plough, the most remarkable application of machinery to agriculture, in this respect resembles the village-made wheel plough. The plough drawn by steam power in like manner turns the second furrow side by side into the first, always throwing the earth the same way, and leaving the ground level. This is one of its defects on heavy, wet land, as it does not drain the surface. But upon the slopes of the Downs no drains or raised "lands" are needed, and the wheel plough answers perfectly.

So perfectly, indeed, does it answer that no iron plough has yet been invented that can beat it, and while the valleys and plains are now almost wholly worked with factory-made ploughs, the South Downs are cultivated with the ploughs made in the villages by the wheelwrights. A wheelwright is generally regularly employed by two or three farms, which keep him in constant work. There is not, perhaps, another home-made implement of old English agriculture left in use; certainly, none at once so curious and interesting, and, when drawn by oxen, so thoroughly characteristic.

Under the September sun, flowers may still be found in sheltered places, as at the side of furze, on the highest of the Downs. Wild thyme continues to bloom—the shepherd's thyme—wild mignonette, blue scabious, white dropwort, yellow bedstraw, and the large purple blooms of greater knapweed. Here and there a blue field gentian is still in flower; "eggs and bacon" grow beside the waggon tracks. Grasshoppers hop among the short dry grass; bees and humble-bees are buzzing about, and there are places quite bright with yellow hawkweeds.

The furze is everywhere full of finches, troops of them; and there are many more swallows than were flying here a month since. No doubt they are on their way southwards, and stay, as it were, on the edge of the sea while yet the sun shines. As the evening falls the sheep come slowly home to the fold. When the flock is penned some stand panting, and the whole body at each pant moves to and fro lengthways; some press against the flakes till the wood creaks; some paw the dry and crumbling ground (arable), making a hollow in which to lie down.

Rooks are fond of the places where sheep have been folded, and perhaps that is one of the causes why they so continually visit certain spots in particular fields to the neglect of the rest.

THE BREEZE ON BEACHY HEAD

The waves coming round the promontory before the west wind still give the idea of a flowing stream, as they did in Homer's days. Here beneath the cliff, standing where beach and sand meet, it is still; the wind passes six hundred feet overhead. But yonder, every larger wave rolling before the breeze breaks over the rocks; a white line of spray rushes along them, gleaming in the sunshine; for a moment the dark rock-wall disappears, till the spray sinks.

The sea seems higher than the spot where I stand, its surface on a higher level—raised like a green mound—as if it could burst in and occupy the space up to the foot of the cliff in a moment. It will not do so, I know; but there is an infinite possibility about the sea; it may do what it is not recorded to have done. It is not to be ordered, it may overleap the bounds human observation has fixed for it. It has a potency unfathomable. There is still something in it not quite grasped and understood—something still to be discovered—a mystery.

So the white spray rushes along the low broken wall of rocks, the sun gleams on the flying fragments of the wave, again it sinks and the rhythmic motion holds the mind, as an invisible force holds back the tide. A faith of expectancy, a sense that something may drift up from the unknown, a large belief in the unseen resources of the endless space out yonder, soothes the mind with dreamy hope.

The little rules and little experiences, all the petty ways of narrow life, are shut off behind by the ponderous and impassable cliff; as if we had dwelt in the dim light of a cave, but coming out at last to look at the sun, a great stone had fallen and closed the entrance, so that there was no return to the shadow. The impassable precipice shuts off our former selves of yesterday, forcing us to look out over the sea only, or up to the deeper heaven.

These breadths draw out the soul; we feel that we have wider thoughts than we knew; the soul has been living, as it were, in a nutshell, all unaware of its own power, and now suddenly finds freedom in the sun and the sky. Straight, as if sawn down from turf to beach, the cliff shuts off the human world, for the sea knows no time and no era; you cannot tell what century it is from the face of the sea. A Roman trireme suddenly rounding the white edge-line of chalk, borne on wind and oar from the Isle of Wight towards the gray castle at Pevensey (already old in older days), would not seem strange. What wonder could surprise us coming from the wonderful sea?

The little rills winding through the sand have made an islet of a detached rock by the beach; limpets cover it, adhering like rivet-heads. In the stillness here, under the roof of the wind so high above, the sound of the sand draining itself is audible. From the cliff blocks of chalk have fallen, leaving hollows as when a knot drops from a beam. They lie crushed together at the base, and on the point of this jagged ridge a wheatear perches.

There are ledges three hundred feet above, and from these now and then a jackdaw glides out and returns again to his place, where, when still and with folded wings, he is but a speck of black. A spire of chalk still higher stands out from the wall, but the rains have got behind it and will cut the crevice deeper and deeper into its foundation. Water, too, has carried the soil from under the turf at the summit over the verge, forming brown streaks.

Upon the beach lies a piece of timber, part of a wreck; the wood is torn and the fibres rent where it was battered against the dull edge of the rocks. The heat of the sun burns, thrown back by the dazzling chalk; the river of ocean flows ceaselessly, casting the spray over the stones; the unchanged sky is blue.

Let us go back and mount the steps at the Gap, and rest on the sward there. I feel that I want the presence of grass. The sky is a softer blue, and the sun genial now the eye and the mind alike are relieved—the one of the strain of too great solitude (not the solitude of the woods), the other of too brilliant and hard a contrast of colours. Touch but the grass and the harmony returns; it is repose after exaltation.

A vessel comes round the promontory; it is not a trireme of old Rome, nor the "fair and stately galley" Count Arnaldus hailed with its seamen singing the mystery of the sea. It is but a brig in ballast, high out of the water, black of hull and dingy of sail: still it is a ship, and there is always an interest about a ship. She is so near, running along but just outside the reef, that the deck is visible. Up rises her stern as the billows come fast and roll under; then her bow lifts, and immediately she rolls, and, loosely swaying with the sea, drives along.

The slope of the billow now behind her is white with the bubbles of her passage, rising, too, from her rudder. Steering athwart with a widening angle from the land, she is laid to clear the distant point of Dungeness. Next, a steamer glides forth, unseen till she passed the cliff; and thus each vessel that comes from the westward has the charm of the unexpected. Eastward there is many a sail working slowly into the wind, and as they approach, talking in the language of flags with the watch on the summit of the Head.

Once now and then the great *Orient* pauses on her outward route to Australia, slowing her engines: the immense length of her hull contains every adjunct of modern life; science, skill, and civilisation are there. She starts, and is lost sight of round the cliff, gone straight away for the very ends of the world. The incident is forgotten, when one morning, as you turn over the newspaper, there is the *Orient* announced to start again. It is like a tale of enchantment; it seems but yesterday that the Head hid her from view; you have scarcely moved, attending to the daily routine of life, and scarce recognise that time has passed at all. In so few hours has the earth been encompassed.

187

The sea-gulls as they settle on the surface ride high out of the water, like the mediaeval caravals, with their sterns almost as tall as the masts. Their unconcerned flight, with crooked wings unbent, as if it were no matter to them whether they flew or floated, in its peculiar jerking motion somewhat reminds one of the lapwing—the heron has it, too, a little—as if aquatic or water-side birds had a common and distinct action of the wing.

Sometimes a porpoise comes along, but just beyond the reef; looking down on him from the verge of the cliff, his course can be watched. His dark body, wet and oily, appears on the surface for two seconds; and then, throwing up his tail like the fluke of an anchor, down he goes. Now look forward, along the waves, some fifty yards or so, and he will come up, the sunshine gleaming on the water as it runs off his back, to again dive, and reappear after a similar interval. Even when the eye can no longer distinguish the form, the spot where he rises is visible, from the slight change in the surface.

The hill receding in hollows leaves a narrow plain between the foot of the sward and the cliff; it is ploughed, and the teams come to the footpath which follows the edge; and thus those who plough the sea and those who plough the land look upon each other. The one sees the vessel change her tack, the other notes the plough turning at the end of the furrow. Bramble bushes project over the dangerous wall of chalk, and grasses fill up the interstices, a hedge suspended in air; but be careful not to reach too far for the blackberries.

The green sea is on the one hand, the yellow stubble on the other. The porpoise dives along beneath, the sheep graze above. Green seaweed lines the reef over which the white spray flies, blue lucerne dots the field. The pebbles of the beach seen from the height mingle in a faint blue tint, as if the distance ground them into coloured sand. Leaving the footpath now, and crossing the stubble to "France," as the wide open hollow in the down is called by the shepherds, it is no easy matter in dry summer weather to climb the steep turf to the furze line above.

Dry grass is as slippery as if it were hair, and the sheep have fed it too close for a grip of the hand. Under the furze (still far from the summit) they have worn a path—a narrow ledge, cut by their cloven feet—through the sward. It is time to rest; and already, looking back, the sea has extended to an indefinite horizon. This climb of a few hundred feet opens a view of so many miles more. But the ships lose their individuality and human character; they are so far, so very far, away, they do not take hold of the sympathies; they seem like sketches—cunningly executed, but only sketches—on the immense canvas of the ocean. There is something unreal about them.

On a calm day, when the surface is smooth as if the brimming ocean had been straked—the rod passed across the top of the measure, thrusting off the irregularities of wave; when the distant green from long simmering under the sun becomes pale; when the sky, without cloud, but with some slight haze in it, likewise loses its hue, and the two so commingle in the pallor of heat that they cannot be separated—then the still ships appear suspended in space. They are as much held from above as upborne from beneath.

They are motionless, midway in space—whether it is sea or air is not to be known. They neither float nor fly; they are suspended. There is no force in the flat sail, the mast is lifeless, the hull without impetus. For hours they linger, changeless as the constellations, still, silent, motionless, phantom vessels on a void sea.

Another climb up from the sheep path, and it is not far then to the terrible edge of that tremendous cliff which rises straighter than a ship's side out of the sea, six hundred feet above the detached rock below, where the limpets cling like rivet heads, and the sand rills run around it. But it is not possible to look down to it—the glance of necessity falls outwards, as a raindrop from the eaves is deflected by the wind, because it *is* the edge where the mould crumbles; the rootlets of the grass are exposed; the chalk is about to break away in flakes.

You cannot lean over as over a parapet, lest such a flake should detach itself—lest a mere trifle should begin to fall, awakening a dread and dormant inclination to slide and finally plunge like it. Stand back; the sea there goes out and out, to the left and to the right, and how far is it to the blue overhead? The eye must stay here a long period, and drink in these distances, before it can adjust the measure, and know exactly what it sees.

The vastness conceals itself, giving us no landmark or milestone. The fleck of cloud yonder, does it part it in two, or is it but a third of the way? The world is an immense cauldron, the ocean fills it, and we are merely on the rim—this narrow land is but a ribbon to the limitlessness yonder. The wind rushes out upon it with wild joy; springing from the edge of the earth, it leaps out over the ocean. Let us go back a few steps and recline on the warm dry turf.

It is pleasant to look back upon the green slope and the hollows and narrow ridges, with sheep and stubble and some low hedges, and oxen and that old, old sloth—the plough—creeping in his path. The sun is bright on the stubble and the corners of furze; there are bees humming yonder, no doubt, and flowers, and hares crouching—the dew dried from around them long since, and waiting for it to fall again; partridges, too, corn-ricks, and the roof of a farmhouse by them. Lit with sunlight are the fields, warm autumn garnering all that is dear to the heart of man, blue heaven above—how sweet the wind comes from these!—the sweeter for the knowledge of the profound abyss behind.

Here, reclining on the grass—the verge of the cliff rising a little, shuts out the actual sea—the glance goes forth into the hollow unsupported. It is sweeter towards the corn-ricks, and yet the mind will not be satisfied but ever turns to the unknown. The edge and the abyss recall us; the boundless plain, for it appears solid as the waves are levelled by distance, demands the gaze. But with use it becomes easier, and the eye labours less. There is a promontory standing out from the main wall, whence you can see the side of the cliff, getting a flank view, as from a tower.

The jackdaws occasionally floating out from the ledge are as mere specks from above, as they were from below. The reef running out from the beach, though now covered by the tide, is visible as you look down on it through the water; the seaweed, which lay matted and half dry on the rocks, is now under the wave. Boats have come round, and are beached; how helplessly little they seem beneath the cliff by the sea!

On returning homewards towards Eastbourne stay awhile by the tumulus on the slope. There are others hidden among the furze; butterflies flutter over them, and the bees hum round by day; by night the nighthawk passes, coming up from the fields and even skirting the sheds and houses below. The rains beat on them, and the storm drives the dead leaves over their low green domes; the waves boom on the shore far down.

How many times has the morning star shone yonder in the East? All the mystery of the sun and of the stars centres around these lowly mounds.

But the glory of these glorious Downs is the breeze. The air in the valleys immediately beneath them is pure and pleasant; but the least climb, even a hundred feet, puts you on a plane with the atmosphere itself, uninterrupted by so much as the tree-tops. It is air without admixture. If it comes from the south, the waves refine it; if inland, the wheat and flowers and grass distil it. The great headland and the whole rib of the promontory is wind-swept and washed with air; the billows of the atmosphere roll over it.

The sun searches out every crevice amongst the grass, nor is there the smallest fragment of surface which is not sweetened by air and light. Underneath, the chalk itself is pure, and the turf thus washed by wind and rain, sun-dried and dew-scented, is a couch prepared with thyme to rest on. Discover some excuse to be up there always, to search for stray mushrooms—they will be stray, for the crop is gathered extremely early in the morning—or to make a list of flowers and grasses; to do anything, and, if not, go always without any pretext. Lands of gold have been found, and lands of spices and precious merchandise; but this is the land of health.

There is the sea below to bathe in, the air of the sky up hither to breathe, the sun to infuse the invisible magnetism of his beams. These are the three potent medicines of nature, and they are medicines that by degrees strengthen not only the body but the unquiet mind. It is not necessary to always look out over the sea. By strolling along the slopes of the ridge a little way inland there is another scene where hills roll on after hills till the last and largest hides those that succeed behind it.

Vast cloud-shadows darken one, and lift their veil from another; like the sea, their tint varies with the hue of the sky over them. Deep narrow valleys—lanes in the hills—draw the footsteps downwards into their solitude, but there is always the delicious air, turn whither you will, and there is always the grass, the touch of which refreshes. Though not in sight, it is pleasant to know that the sea is close at hand, and that you have only to mount to the ridge to view it. At sunset the curves of the shore westward are filled with a luminous mist.

Or if it should be calm, and you should like to look at the massive headland from the level of the sea, row out a mile from the beach. Eastwards a bank of red vapour shuts in the sea, the wavelets—no larger than those raised by the oar—on that side are purple as if wine had been spilt upon them, but westwards the ripples shimmer with palest gold.

The sun sinks behind the summit of the Downs, and slender streaks of purple are drawn along above them. A shadow comes forth from the cliff; a duskiness dwells on the water; something tempts the eye upwards, and near the zenith there is a star.

END

192